Renewable Energy: Research, Development and Policies

RENEWABLE ENERGY GRID INTEGRATION

THE BUSINESS OF PHOTOVOLTAICS

MARCO H. BALDERAS

EDITOR

NOVA

Nova Science Publishers, Inc.

New York

For permission to use material from this book please contact us:
Telephone 631-231-7269; Fax 631-231-8175
Web Site: http://www.novapublishers.com

NOTICE TO THE READER
The Publisher has taken reasonable care in the preparation of this book, but makes no expressed or implied warranty of any kind and assumes no responsibility for any errors or omissions. No liability is assumed for incidental or consequential damages in connection with or arising out of information contained in this book. The Publisher shall not be liable for any special, consequential, or exemplary damages resulting, in whole or in part, from the readers' use of, or reliance upon, this material.

Independent verification should be sought for any data, advice or recommendations contained in this book. In addition, no responsibility is assumed by the publisher for any injury and/or damage to persons or property arising from any methods, products, instructions, ideas or otherwise contained in this publication.

This publication is designed to provide accurate and authoritative information with regard to the subject matter covered herein. It is sold with the clear understanding that the Publisher is not engaged in rendering legal or any other professional services. If legal or any other expert assistance is required, the services of a competent person should be sought. FROM A DECLARATION OF PARTICIPANTS JOINTLY ADOPTED BY A COMMITTEE OF THE AMERICAN BAR ASSOCIATION AND A COMMITTEE OF PUBLISHERS.

LIBRARY OF CONGRESS CATALOGING-IN-PUBLICATION DATA

Renewable energy grid integration : the business of photovoltaics / Editor, Marco H. Balderas.
257 p. : ill. ; 27cm.
Includes bibliographical references and index.
ISBN: 978-1-60741-324-0 (hardcover)
1. Building-integrated photovoltaic systems. 2.Photovoltaic power generation. 3.Photovoltaic power systems. 4.Photovoltatic cells. I. Baldera, Marco H.
TK1087 .R47 2009
(OCoLC)ocn316825443

2009289613

Published by Nova Science Publishers, Inc. †*New York*

RENEWABLE ENERGY GRID INTEGRATION

THE BUSINESS OF PHOTOVOLTAICS

RENEWABLE ENERGY: RESEARCH, DEVELOPMENT AND POLICIES

Ethanol and Biofuels: Production, Standards and Potential
Wesley P. Leland
2009 ISBN 978-1-60692-224-8

Renewable Energy Grid Integration: The Business of Photovoltaics
Marco H. Balderas
2009 ISBN 978-1-60741-324-0

CONTENTS

PREFACE

Now is the time to plan for the integration of significant quantities of distributed renewable energy into the electricity grid. Concerns about climate change, the adoption of state-level renewable portfolio standards and incentives, and accelerated cost reductions are driving steep growth in U.S. renewable energy technologies. This book addresses such concerns and in particular, the number of distributed solar photovoltaic (PV) installations, which is growing very rapidly. As distributed PV and other renewable energy technologies mature, they can provide a significant share of the nation's electricity demand. However, as their market share grows, concerns about potential impacts on the stability and operation of the electricity grid may create barriers to their future expansion. To facilitate more extensive adoption of renewable distributed electric generation, this book discusses the the Renewable Systems Interconnection (RSI) study, which looks at the technical and analytical challenges that must be addressed to enable high penetration levels of distributed renewable energy technologies.

In: Renewable Energy Grid Integration
Editor: Marco H. Balderas

ISBN: 978-1-60741-324-0
© 2009 Nova Science Publishers, Inc.

Chapter 1

ROOFTOP PHOTOVOLTAICS MARKET PENETRATION SCENARIOS[*]

*J. Paidipati, L. Frantzis,
H. Sawyer and A. Kurrasch*

PREFACE

Now is the time to plan for the integration of significant quantities of distributed renewable energy into the electricity grid. Concerns about climate change, the adoption of state-level renewable portfolio standards and incentives, and accelerated cost reductions are driving steep growth in U.S. renewable energy technologies. The number of distributed solar photovoltaic (PV) installations, in particular, is growing rapidly. As distributed PV and other renewable energy technologies mature, they can provide a significant share of our nation's electricity demand. However, as their market share grows, concerns about potential impacts on the stability and operation of the electricity grid may create barriers to their future expansion.

To facilitate more extensive adoption of renewable distributed electric generation, the U.S. Department of Energy launched the Renewable Systems Interconnection (RSI) study during the spring of 2007. This study addresses the technical and analytical challenges that must be addressed to enable high penetration levels of distributed renewable energy technologies. Because integration-related issues at the distribution system are likely to emerge first for PV technology, the RSI study focuses on this area. A key goal of the RSI study is to identify the research and development needed to build the foundation for a high-penetration renewable energy future while enhancing the operation of the electricity grid.

The RSI study consists of 15 reports that address a variety of issues related to distributed systems technology development; advanced distribution systems integration; system-level tests and demonstrations; technical and market analysis; resource assessment; and codes, standards, and regulatory implementation. The RSI reports are:

---

[*] Excerpted from Subcontract Report NREL/SR-581-42306, dated February 2008.

- Renewable Systems Interconnection: Executive Summary
- Distributed Photovoltaic Systems Design and Technology Requirements
- Advanced Grid Planning and Operation
- Utility Models, Analysis, and Simulation Tools
- Cyber Security Analysis
- Power System Planning: Emerging Practices Suitable for Evaluating the Impact of High-Penetration Photovoltaics
- Distribution System Voltage Performance Analysis for High-Penetration Photovoltaics
- Enhanced Reliability of Photovoltaic Systems with Energy Storage and Controls
- Transmission System Performance Analysis for High-Penetration Photovoltaics
- Solar Resource Assessment
- Test and Demonstration Program Definition
- Photovoltaics Value Analysis
- Photovoltaics Business Models
- Production Cost Modeling for High Levels of Photovoltaic Penetration
- Rooftop Photovoltaics Market Penetration Scenarios.

Addressing grid-integration issues is a necessary prerequisite for the long-term viability of the distributed renewable energy industry, in general, and the distributed PV industry, in particular. The RSI study is one step on this path. The Department of Energy is also working with stakeholders to develop a research and development plan aimed at making this vision a reality.

ACKNOWLEDGMENTS

Navigant Consulting Inc. (NCI) would like to first thank the U.S. Department of Energy (DOE) and the DOE National Renewable Energy Laboratory (NREL) for sponsoring this work. NCI would also like to thank the McGraw-Hill Companies for providing floor space data, a crucial component of this study. Furthermore, many independent reviewers took the time to review this chapter and provide valuable insights; their efforts were very much appreciated.

LIST OF ACRONYMS

BAU	business as usual scenario
CBECS	Commercial Building Energy Consumption Survey
DOE	U.S. Department of Energy
EIA	Energy Information Administration
FERC	Federal Energy Regulatory Commission
IREC	Interstate Renewable Energy Council
MW	megawatt
MWh	megawatt-hour
NCI	Navigant Consulting, Inc.
NREL	National Renewable Energy Laboratory

O&M	operation and maintenance
RECS	Residential Energy Consumption Survey
RPS	Renewable Portfolio Standard
RSI	Renewable System Integration (study)
SAI	Solar America Initiative
TOU	time of use

EXECUTIVE SUMMARY

The goal of this study was to model the market penetration of rooftop photovoltaics (PV) in the United States under a variety of scenarios, on a state-by-state basis, from 2007 to 2015. The study was performed by Navigant Consulting Inc. (NCI) for the U.S. Department of Energy (DOE) under a subcontract to the DOE National Renewable Energy Laboratory. The model looked at the retrofit and new construction segments of the residential and commercial rooftop markets. For each state, the model calculated the market penetration percent, annual installations, and cumulative installations. The scenarios studied included net metering rules, electric rate tariff levels and structures, the availability of financial incentives, system pricing, and carbon legislation.

To perform the market penetration analysis, NCI first calculated the technical potential for PV implementation for each of the 50 states by using data on floor space, building characteristics, PV solar access factors, and PV system efficiency. Next, based on a selection of 98 representative utilities within the states and the District of Columbia, NCI calculated economic potential using current electric rate structures and tariffs, local and federal incentive levels, system costs, operation and maintenance (O&M) and inverter replacement costs, building load profiles, PV output profiles, and net metering rules. This work yielded a simple payback period, which was incorporated into a market penetration curve. To arrive at the final estimate of economic potential, the market penetration results were augmented by a technology adoption curve, screens related to interconnection standards, and Renewable Portfolio Standard (RPS) solar set-aside requirements.

NCI ran a variety of scenarios to examine the impacts of different variables, including variations on system pricing, interconnection standards, net metering availability, net metering caps, carbon legislation, electric price escalation, availability of time-of-use rates, RPS enforcement, and availability of federal and local incentives for PV. The variables with the largest impact on market penetration were system pricing, net metering policy, extending the commercial and residential federal tax credits to 2015 (as opposed to our baseline assumption of commercial incentives to 2015 and residential ones to 2010), and interconnection policy, as shown in figure E- 1.

Figure E- 1 illustrates that there is significant potential in the United States for PV on buildings. However, several variables that were not modeled in this study could impact the results. Constraints along the PV supply chain (such as the current silicon shortage) could result in higher module prices or constrained supply, thus decreasing market penetration. In addition, significant international demand could draw supply away from the U.S. market, thus decreasing U.S. market penetration. In contrast, new state or federal policies, such as incentive programs or RPS, could drive U.S. demand even higher.

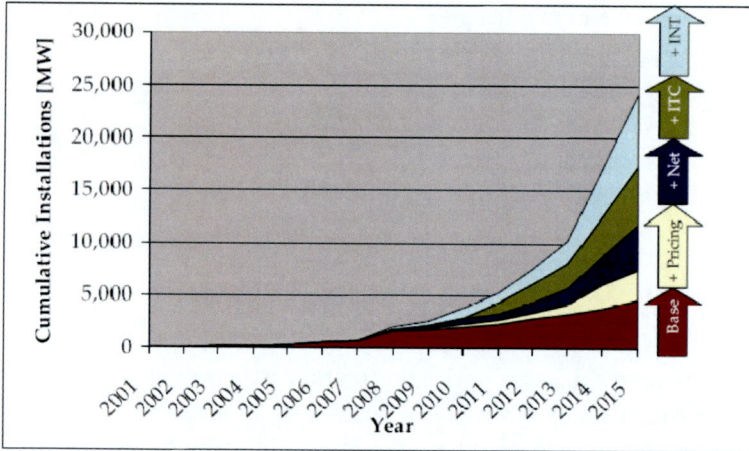

Figure E-1. Influence on cumulative U.S. PV installations of system pricing, net metering policy, federal tax credits, and interconnection standards.

1.0. INTRODUCTION

The economic viability of photovoltaics (PV) in the United States is a function of several variables, including electricity prices, system costs, net metering laws, and incentives. Given the fragmented nature of electricity markets, regulations, and incentives, the economics of PV need to be assessed locally. Accordingly, for this study, we modeled the market penetration of rooftop PV in the United States under a variety of scenarios, on a state-by-state basis, from 2007 to 2015.

The study was performed by Navigant Consulting Inc. (NCI) for the U.S. Department of Energy (DOE) under a subcontract to the DOE National Renewable Energy Laboratory (NREL). The analysts were challenged to ensure that the modeling methodology was highly clear and transparent. The model looked at the retrofit and new construction segments of the residential and commercial rooftop markets. It did not include field-based systems, a potentially significant market segment for growth. It also did not capture price dynamics related to international competition for PV modules, or downward changes in electricity prices resulting from a potential drop in demand because of PV.

For each state, the model calculated the percent market penetration, annual installations, and cumulative installations. The scenarios studied included net metering rules, electric rate tariff levels and structures, availability of financial incentives, system pricing, and carbon legislation. This chapter and the current version of the model are important early steps in the development of a better understanding of the market dynamics of the U.S. PV industry.

2.0. CURRENT STATUS OF THE RESEARCH

Many market studies of the PV industry have been performed during the past few years. Examples include DOE PV road maps (www.eere.energy.gov/solar/deployment.html), PV Services Program reports (www.navigantconsulting.com), Solarbuzz projections (www.solarbuzz.com), and reports from the Prometheus Institute (www.prometheus.com). NCI and others have completed in-depth market penetration studies for constrained areas (Arizona, California, and Austin, Texas), but each of these markets is unique, so study results cannot be extrapolated to the entire nation.

Most previous studies have not used a market penetration approach that captures all facets of project economics. Prior projections have used a variety of approaches:

- A simple extrapolation of historical PV demand, using factors to represent aggressive or decreasing demand
- Market surveys to obtain key player views on future projections
- Reviews of the projected levelized cost of electricity for PV versus retail electricity rates to assess project attractiveness.

None of these methods, however, are in publicly available models. The goal of this research was to create a publicly available model that captures local variables such as retail electric rates, insolation levels, weather (and hence building load), incentives, net metering policy, and interconnection policy.

3.0. PROJECT APPROACH

NCI created a Microsoft Excel©-based spreadsheet tool for calculating market penetration. shows a flow diagram of the model. This chapter discusses each section of the model: technical potential, economic potential, and the scenarios studied.

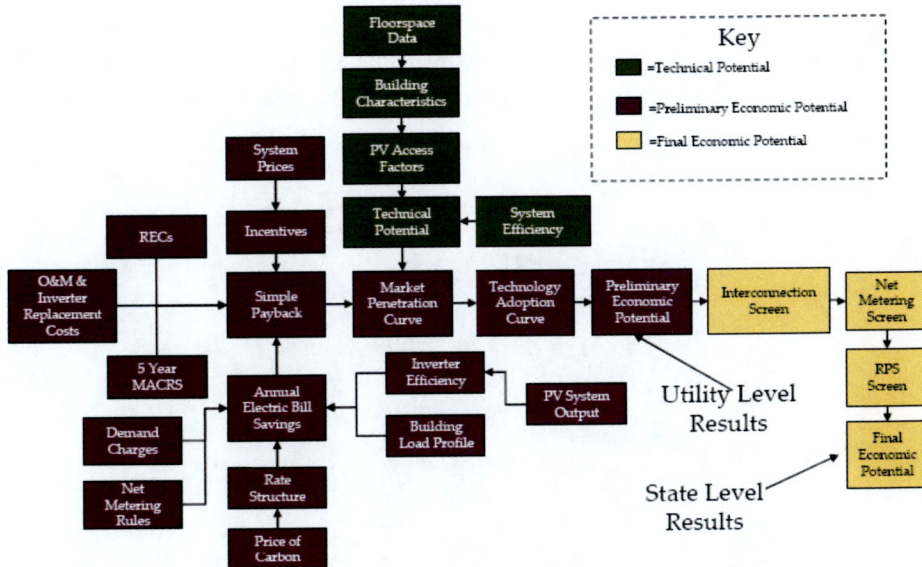

Figure 1. Market penetration flow diagram.

3.1. Technical Potential

To calculate the market penetration of PV, we must first know the size of the available market. Current and projected total U.S. roof space was thus estimated for 2007 through 2015, by state, for residential and commercial buildings. A PV solar access factor was then applied to the roof space data to estimate how much roof space is actually available for PV. The PV access factor takes into account shading, building orientation, and roof structural soundness. PV power density data are then used to calculate potential installed capacity on a state-by-state basis.

To calculate total roof space, we began with data on the total amount of floor space in residential and commercial buildings, by state, from McGraw-Hill for 2007 through 2011. They used the growth (or decline) trends from 2007 to 2011 to project growth (or decline) from 2012 to 2015. To estimate how floor space translates into roof space, we used data on the average number of floors per building from the Energy Information Administration's (EIA) Residential Energy Consumption Survey (RECS) and Commercial Building Energy Consumption Survey (CBECS) databases. For pitched roofs, assumed to be 92% of the residential market, NCI assumed an 18-degree pitch to calculate roof space. Although 18 degrees is a typical number, the angle can very from 0 to 45 degrees in any given region. We defined new construction based upon the floor space added in any year.

To estimate how much of the total roof space is available for PV, NCI developed PV access factors based on a study for a major U.S. utility company. The study was adjusted for California conditions after interviews with Ed Kern of Irradiance, who has many years of installation experience in the industry. Separate access factors were developed for cooler and warmer climates. State designations are shown in figure 2. Figure 3, 4, 5, and 6 show the different analyses with the assumptions used for flat residential roofs. The PV access factors were then applied to state-level roof space data to estimate the available roof area for PV. The results should not be confused with the share of homes that are not suitable for PV, however, since the study is focusing on roof space. However, the factors used in the study (~25% for residential and ~60% for commercial) are similar to the space taken up by current PV systems.

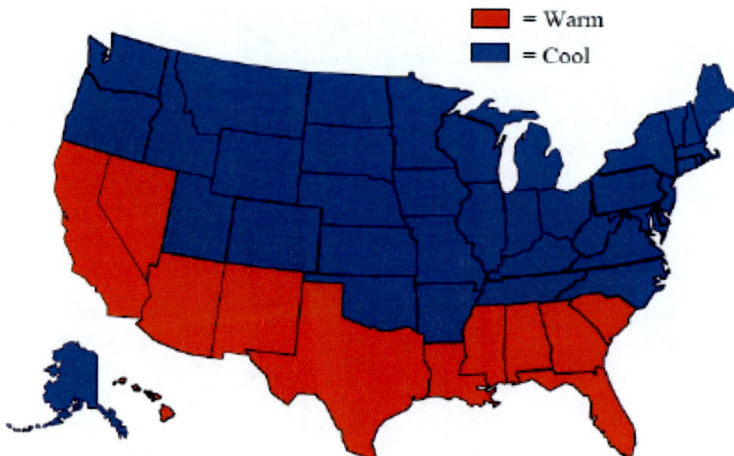

Figure 2. State-level climate type designations.

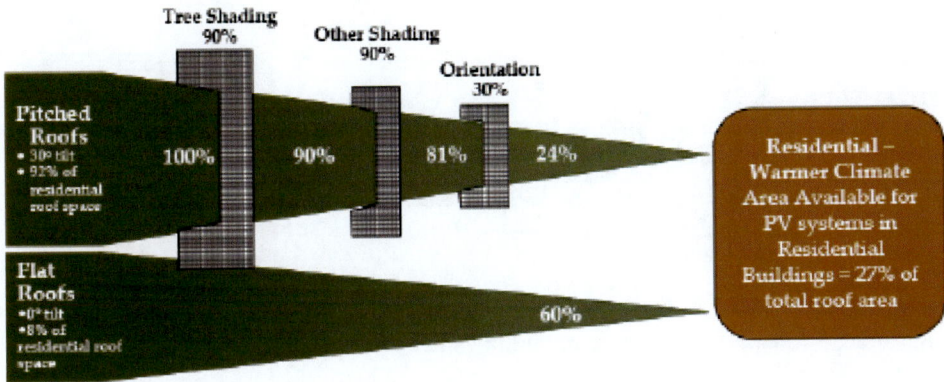

Figure 3. PV access factor for residential buildings in warmer climates.

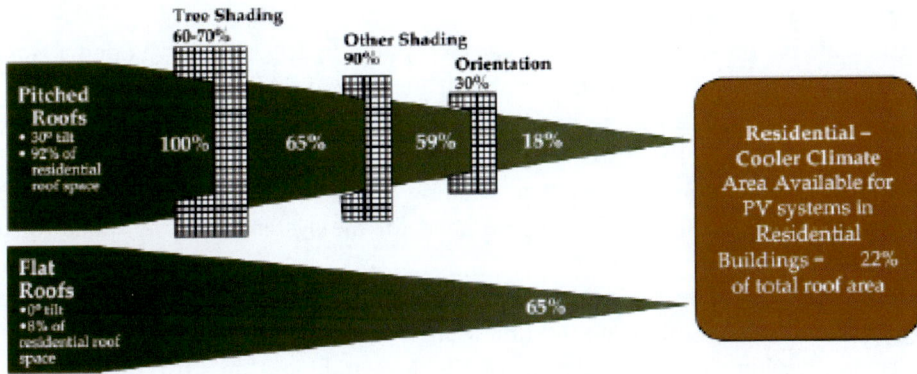

Figure 4. PV access factor for residential buildings in cooler climates.

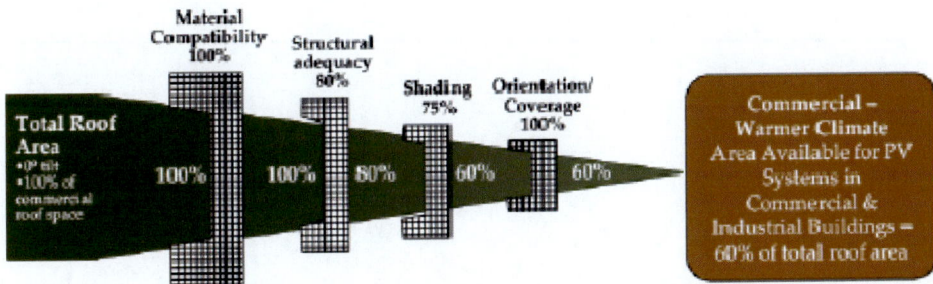

Figure 5. PV access factor for commercial buildings in warmer climates.

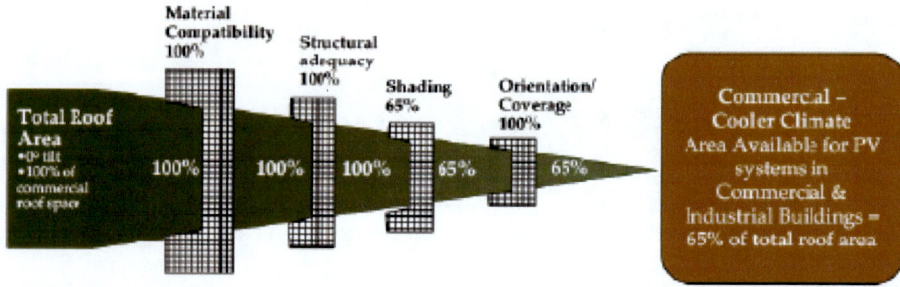

Figure 6. PV access factor for commercial buildings in cooler climates.

We estimated the technical potential using data on PV power density from DOE's Solar America Initiative Technology Pathway Partnership (for information, see www.eere.energy.gov/solar/ solar_america/index.html). Technical potential is defined as PV system power density (in MWpDC per million square feet) times the roof space available for PV in a given area.

To calculate the power density of a solar PV system in 2007, we developed a weighted-average module efficiency using market share for the three most prevalent technologies in the market today. The power density of a module was then calculated on a square-footage basis, and the power density of a PV system was calculated by applying a packing factor of 1.25 for residential and commercial systems. The packing factor modifies (as a decrease) the PV power density by taking into account space need for the system, such as space for access between modules, wiring, and inverters.

The resulting system power density is 10 MW/million ft^2, as derived from an average module efficiency of 13.5%. For 2015, we assumed an average module efficiency of 18.5% for all installations, resulting in a power density of 13.7 MW/million ft^2 in 2015. Figure 7 shows the technical potential in 2015. Technical potential increases over time for two reasons: rooftop area grows over time and system efficiencies increase over time. See the appendix in this chapter for a table of state-by-state results.

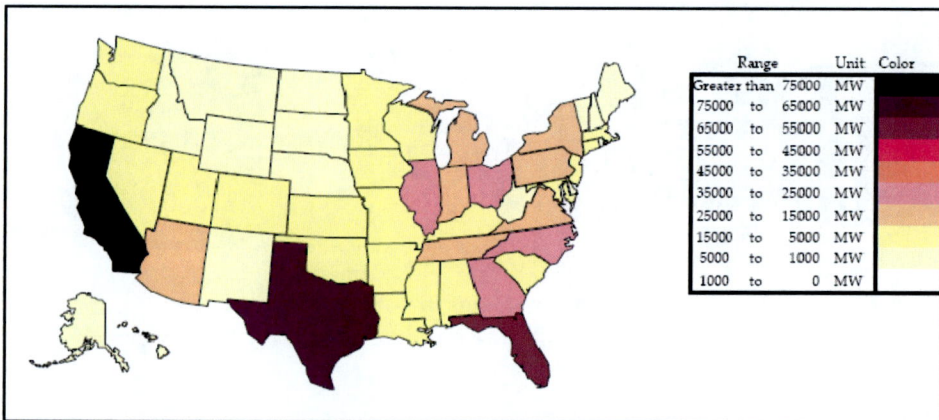

Figure 7. U.S. rooftop PV technical potential in 2015 (independent of economics).

3.2. Preliminary Economic Potential

After calculating the technical potential for each state, we looked at the economics of PV to assess the economic potential. Referring back to figure 3-1, economic potential is calculated by taking market penetration as a percentage of technical potential and multiplying the results by a technology adoption curve.

The input to NCI's market penetration curves is simple payback, so we picked from one to five utilities in each state to represent PV economics. For each utility analyzed (or state, for certain variables), we collected rate structure and tariff data, net metering rules, incentives data, building load profiles, and PV output profiles. See the appendix for more details about the sources and values of each of these variables and the list of utilities analyzed, by state.

Equation 1 shows the simple payback calculation for the residential market, and Equation 2 shows the calculation for the commercial market. Note that, according to EIA's CBECS database, approximately 25% of all commercial building floor space is contained in buildings that do not pay taxes (such as schools and government buildings), so this calculation is somewhat conservative for those segments.

$$\text{Simple Payback} = \frac{[\text{Installed Cost} - \text{Federal Incentives} - \text{Capacity Based Incentives} + \text{Tax Rate} * \text{Rebate Amount}]}{[\text{Annual Electric Bill Savings} + \text{Performance Based Incentives} - \text{O\&M Costs}]}$$

Equation 1. Residential simple payback.

$$\text{Simple Payback} = \frac{[\text{Installed Cost} - \text{Federal Incentives} - \text{Capacity Based Incentives} + \text{Tax Rate} * \text{Rebate Amount}]}{[(1 - \text{Tax Rate}) * (\text{Annual Electric Bill Savings} - \text{O\&M Costs}) + \text{Performance Based Incentives} + \text{Amortized MACRS savings}]}$$

Equation 2. Commercial simple payback.

We used two different market penetration curves (both of which use simple payback as inputs): one for the retrofit market and one for the new construction market. Figure 8 shows the market penetration curves used. Based on interviews with key stakeholders, we used a different curve for new construction because builders are in general reluctant to add PV as a standard feature and require shorter paybacks before making it standard. We used two studies of market penetration to develop curves for this study. Kastovich et al. calculated market penetration curves for retrofit and new construction markets of energy technologies. They surveyed customer behaviors based on simple payback. NCI produced a curve based on field interviews, consumer surveys, and market data on the adoption of efficient energy technologies in the market, again based on simple payback.

Several variables could influence the evolution of these market penetration curves over time. The most important would be government policies that support the adoption of PV. One example is the California Solar Initiative, which after 2010 requires that all new subdivisions with more than 50 homes must offer PV as an option to potential homebuyers. Another variable could be consumer awareness campaigns that shift consumer behavior to adopt PV at higher paybacks.

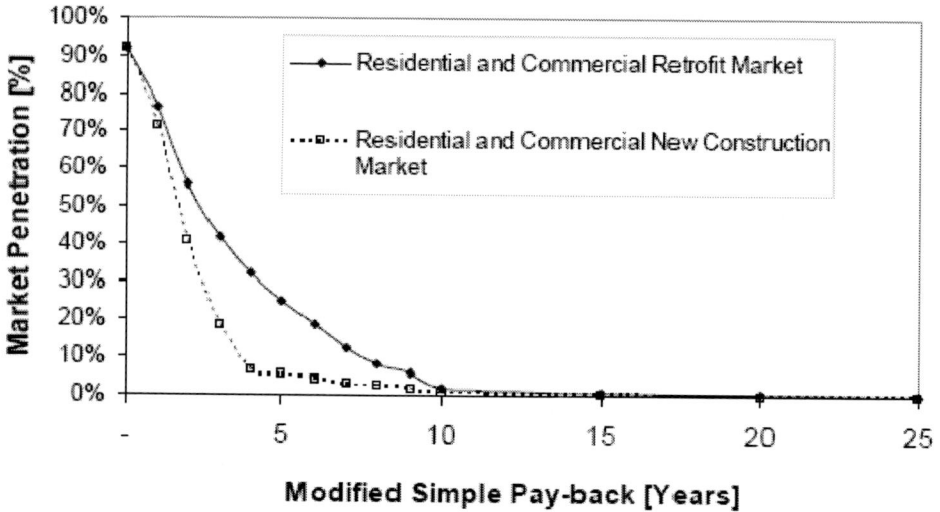

Figure 8. Market penetration curves used.

After calculating the percent market penetration, we used an S-curve to model technology adoption. An S-curve provides the rate of adoption of technologies as a function of the technology's characteristics and market conditions. Figure 9 shows the S-curves used, which are Fisher-Pry curves. The Fisher-Pry technology substitution model predicts the market adoption rate for an existing market of known size. We used this model because consumers are replacing grid power with PV power. The market of known size comes from technical potential and market potential calculations.

The rate at which technologies are adopted depends on several market characteristics: technology characteristics (e.g., technology economics, new vs. retrofit); industry characteristics (e.g., industry growth, competition); and external factors (e.g., government regulation, trade restrictions). Historical data collected by Fisher-Pry and NCI reveal that major classes of technology/segment with common segment-penetration characteristics can be classified into five categories, each with its own time to segment saturation, as shown in table 3-1.

For PV, we picked the two classes that closely resembled the PV market in the United States, class B and class C. They then used the average of the two classes' curves, as shown in Figure 9.

Because 2007 was more than half over when this chapter was written, the model assumes annual installations and cumulative installations through 2007 and starts calculating penetration for 2008.

After applying these curves, we arrived at cumulative installations up to the year of analysis. A final market penetration was calculated after applying the RPS and interconnection screens discussed in the next section. Final market penetration is defined as cumulative installations (defined by peak DC rating) in a given area as a percentage of the technical potential in that area.

Table 1. Five Classes of Technology Adoption Characteristics (Fisher-Pry)

Characteristics					
Time to Saturation (t_s)	5 years	10 years	20 years	40 years	>40 years
Technology Factors					
Equipment Life	< 5 years	5–15 years	15–25 years	25–45 years	>40 years
Equipment Replacement	None	Minor	Unit operation	Plant section	Entire plant
Technology Experience	New to U.S. only	New to U.S. only	New to U.S. only	New	New
Industry Factors					
Growth (% per year)	>5%	>5%	2~5%	1–2%	<1%
Attitude to Risk	Open	Open	Cautious	Conservative	Adverse
External Factors					
Government Regulation	Forcing	Forcing	Driving	None	None

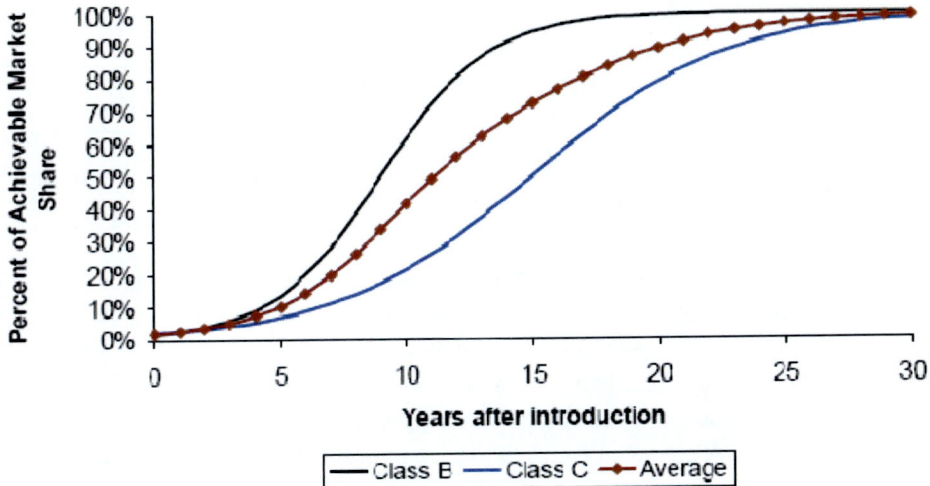

Figure 9. Technology adoption curve used.

3.3. Scenarios Analyzed

We developed a set of scenarios dealing with interconnection policy, RPS solar set-aside policy, system pricing, net metering policy, carbon legislation, rate structure policy, electric rate escalation, and federal incentives. For the first of the scenarios, we used data provided to DOE from the Interstate Renewable Energy Council's (IREC) assessment of each state's interconnection standards (or utility's, in states without state-level laws) in regard to facilitating distributed generation. IREC gave each location a rating on a five-point scale, as shown in table 2, that assesses the likelihood of a system being installed. We then translated these assessments into an assumed percentage of achievable market, also shown in table 2. They scaled preliminary economic potential by this amount. (See the appendix for a complete list of state rankings.)

Many states' interconnection standards are a barrier to the wider adoption of PV, although several are considering revising them. Recognizing this, we created a scenario in which all states improve their interconnection standards to the point at which the standards do not hinder PV interconnection (i.e., a "superior" ranking in IREC's scale in table 2).

Table 2. IREC's Interconnection Assessment Rating System

IREC Rating	IREC's Assessment	NCI's Assumed Achievable Market
Superior (A)	Interconnection policies encourage distributed generation	100%
Good (B)	Interconnection policies contain some difficulties but less than 5% of solar projects will incur needless costs or delays because of interconnection problems	95%
Fair (C)	Interconnection policies allow interconnection but with some difficulty. Up to 25% of proposed solar projects will incur needless delays, costs, or some will fail because of interconnection	75%
Poor (D)	Interconnection policies are very poor. Costs of systems and time to complete interconnection will be significant. Up to 50% of projects will incur significant costs and delays to complete interconnection process. An undesirable number of projects will fail.	60%
Barrier (E)	Interconnection policies represent a major barrier to the use of solar. 50% or greater will experience significant costs, delays or project cancellation because of interconnection policies	40%

Some states or utilities have net metering caps, typically expressed as a percentage of the utility's or state's peak load. This study used EIA peak demand data to translate net metering caps as percentages into megawatts. For each year of analysis, market penetration is the ratio of cumulative installations to net metering caps. The model assumes that if net metering caps are reached in a given year, net metering is not allowed in the next year of analysis. We used EIA's Annual Energy Outlook projections for load growth to estimate how peak demand will change over time.

The next two scenarios concern net metering standards. The first net metering scenario assumes all net metering caps are lifted in 2007. The second one concerns the availability of net metering. Currently, most states and the District of Columbia offer net metering, but some states and utilities still do not allow it. Figure 10 shows net metering assumptions for the utilities used in this study, by state. This scenario assumes net metering is available nationwide, starting in 2008.

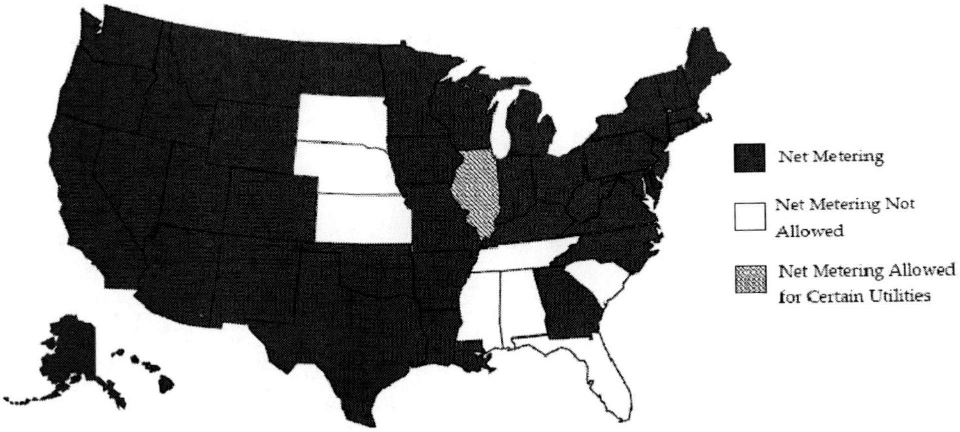

Figure 10. Availability of net metering.

The next scenario involved RPS solar set-asides. Several states have solar set-asides or distributed generation set-asides as part of their RPS (i.e., a certain percentage of RPS megawatt-hours must be from PV systems). For each year of analysis, the market penetration model will ensure that market penetration at least meets the level required by solar set-asides, independent of net metering caps, economics, or poor interconnection standards. The exact mechanisms for this are not specified, but examples could be extra utility rebates or utilities owning rooftop PV systems.

For reference, figure 11 shows solar set-aside requirements in 2015. As shown in the figure, RPS could account for a total of ~2,200 MW of installed PV in 2015. Achieving these goals will depend on a number of factors, such as compliance mechanisms, so they may or may not be met. The model has a switch in which RPS solar set-asides goals are met or not met.

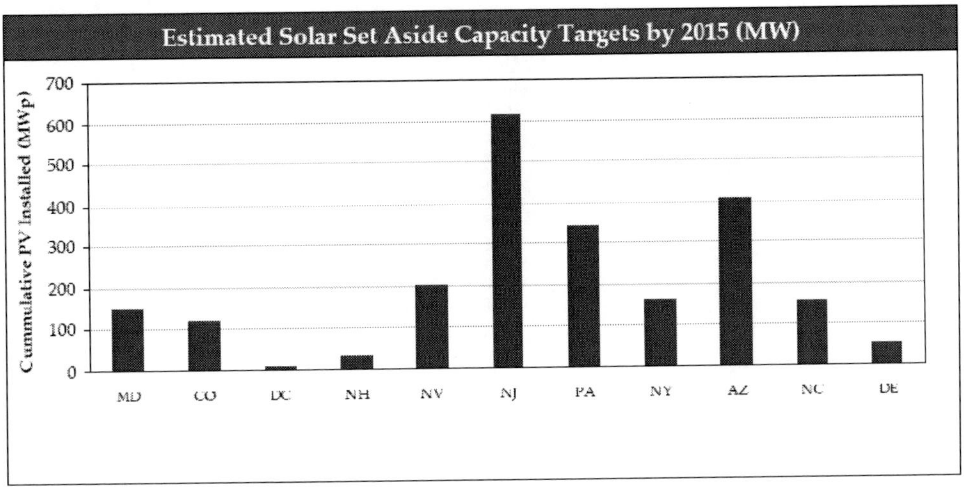

Figure 11. Solar set-aside targets.

NCI used two different system pricing cases. The first case assumed that system prices decline at historical rates. The second case used targets from the DOE's Solar America

Initiative (SAI) program. DOE's targets are based on a combination of internal analysis of potential cost reductions in PV technologies and a review of information provided in applications submitted to the SAI Technology Pathway Partnership solicitation during 2006. table 3 lists the two pricing cases.

Table 3. System Pricing Assumptions

System Price Scenario	Market Segment	Retrofit Installed System Price ($2007/Wpdc)			New Construction Installed System Price ($2007/Wpdc)		
		2007	2010	2015	2007	2010	2015
Business as Usual (BAU)	Residential	$7.40	$6.20	$4.80	$7.40	$5.90	$4.50
	Commercial	$6.41	$5.80	$4.50	$6.70	$5.50	$4.20
Solar America Initiative (SAI)	Residential	$7.40	$5.11	$3.10	$7.10	$3.86	$2.44
	Commercial	$6.41	$3.75	$2.49	$6.23	$3.60	$2.32

At the time of this project, several bills were circulating through the U.S. House of Representatives and the U.S. Senate that would introduce some type of carbon legislation. During the course of this project, for illustration purposes the study used the Senate's Low Carbon Economy Act bill sponsored by Senator Bingaman of New Mexico. The Act creates a national cap-and-trade system with a ceiling on the price of carbon, as shown in table 4. We assume that carbon will trade at the ceiling price. To assess the effect of this on potential PV customers, we used carbon intensity data from EIA (in tonnes of CO_2 per kWh) and modeled the price of carbon as a surcharge on electric bills. Refer to the appendix for details on the calculations. Thus, we modeled a scenario that assumes the legislation is introduced.

Table 4. Provisions of Low Carbon Economy Act

Year	Ceiling on Carbon Price [$/Tonne CO2]
2007	$0.00
2008	$0.00
2009	$0.00
2010	$0.00
2011	$0.00
2012	$12.00
2013	$12.60
2014	$13.23
2015	$13.89

Time-of-use (TOU) rates can significantly impact PV economics, yet they are not available in all areas. We created a scenario in which TOU rates are made available from every utility. To create TOU rates, we used a rate-multiplier approach. Within the eight North American Electric Reliability Corporation (NERC) regions, utilities from each state with established TOU rates were selected for analysis. For each utility, we calculated the ratio of peak-tostandard and non-peak-to-standard rates for both the summer and winter seasons. Overall averages of those ratios were then taken for each region to use as benchmarks when estimating TOU rates for utilities that do not offer them. Another component of the rate- multiplier analysis involved calculating an average number of peak hours and start times of those peak periods within each region. See the appendix for more detail.

Given the influence of electricity prices on simple payback, we looked at three different forward price projections. The first (and most conservative) projection uses EIA's Annual Energy Outlook pricing projections. These projections show real cost decreases over time. The second projection uses state-by-state projections developed by NCI using NERC reports, ISO reports, and other data sources to look at the impact of policy changes (e.g., rate caps lifted), capacity shortfalls, and market dynamics. The result was an annual percentage year-over-year change in price, by state.

The final two scenarios we analyzed involved federal incentives for PV. Federal residential incentives (tax credits) are set to expire at the end of 2008; at that time, the commercial incentive will be reduced from 30% to 10%. However, the U.S. House of Representatives and the U.S. Senate are working on legislation to extend those tax credits. Each chamber has different provisions for extension, and we worked with the Solar Energy Industries Association to come up with a best estimate about which legislation will pass. The first scenario assumes the commercial incentive is extended to 2015 and the residential incentive is extended to 2010, with the $2,000-per-system cap lifted. The second scenario assumes that both the residential and commercial credits are fully extended to 2015, with the $2,000-persystem cap lifted.

Many participants in the PV market have concerns regarding the availability of installers to meet a growing demand. In discussing this issue with stakeholders, we found that the time to train a qualified PV installer ranges from six weeks to three months, which fits within the one- year temporal resolution of this model. To understand future requirements for installers, we calculated estimated installer requirements state by state for each year of analysis.

4.0. PROJECT RESULTS

We conducted several model runs, varying each of the scenarios. The first run used values for each variable that provided the least support for PV penetration. The next run served as a base case and used inputs that are more representative of what is likely to occur. Next, using the base case as a starting point, we looked at the impact of individual policy improvements for net metering, interconnection standards, and TOU rates, along with a full extension of the residential federal tax credit. Using the results of these four runs, we chose the two variables with the largest impact and looked at the results. Finally, we conducted a best-case run within the context of this model/set of assumptions. There is potential for more rapid market penetration, for example, if electricity prices rise faster then projected here, if states (or the federal government) institute more aggressive solar or climate-related policies, and so on. All runs were done using business-as-usual (BAU) and SAI system pricing.

Table 5. Inputs into Each Run

	Worst-Case	Base-Case	Focused Policies	Best-Case
Interconnection Policy Scenario	Current Rules	Current Rules	Current Rules	Improved
Net Metering Availability Scenario	Current Availability	Current Availability	Nationwide Availability	Nationwide Availability
Net Metering Cap Scenario	Current Caps	Current Caps	Caps Lifted	Caps Lifted
Cap and Trade Scenario	None	Low Carbon Economy Act	Low Carbon Economy Act	Low Carbon Economy Act
Electricity Price Escalation	EIA's Projections	Accelerated	Accelerated	Accelerated
Federal Tax Credit	Baseline	Extended	Fully Extended	Fully Extended
Time-of-Use Rates	Current Availability	Current Availability	Current Availability	Nationwide Availability
RPS Solar Set Aside Enforcement	No	Yes	Yes	Yes

4.1. The Worst Case

The first run used the worst case for each input assumption, as shown in table 6. The run assumed that federal tax credits are not extended, carbon legislation is not passed, system price declines occur at historical rates, and electricity prices evolve per the EIA's projections. All of these factors combine to decrease the economic attractiveness of PV.

Table 6. Worst-Case Scenario Inputs

Scenario	Value
System Pricing Scenario	Business-As-Usual
Interconnection Policy Scenario	Current Rules
Net Metering Availability Scenario	Current Availability
Net Metering Cap Scenario	Current Caps
Cap and Trade Scenario	None
Electricity Price Escalation	EIA's Projections
Federal Tax Credit	Baseline
Time-of-Use Rates	Current Availability
RPS Solar Set-Aside Enforcement	No

Figure 12 shows cumulative installations by state for 2015. See the appendix for a table of state-by-state results. Installations are strong in 2007 and 2008, but once the federal tax credits expire, the market shrinks by 90% in 2009. The only state in which significant installations occur is California, where the California Solar Initiative mitigates the loss of federal tax credits. The assumption that RPS solar set-asides are not enforced has a large impact, as shown in figure 13. Given that most RPS have a ceiling on alternative compliance payments, market forces (i.e., a lucrative renewable energy credit, or REC, price improves system economics) can only go so far in enforcing the solar set-asides.

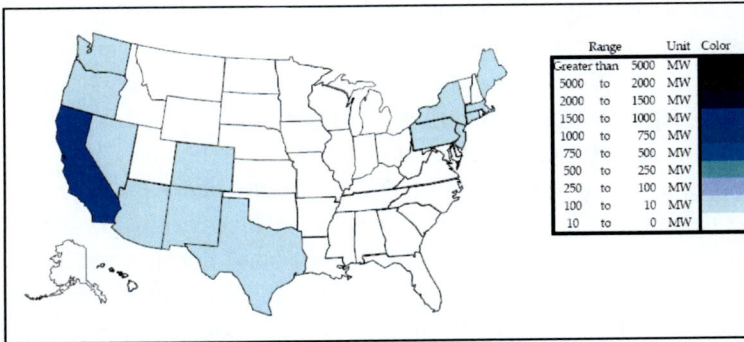

Figure 12. Cumulative installations in 2015 under the worst case.

Table 7. Nationwide Results for the Worst Case

Year	Annual Installations [MW]	Cumulative Installation [MW]	Installers Required [FTE]	Market Penetration [%]
2007	251	733	1,864	0.16%
2008	155	889	1,117	0.19%
2009	35	924	245	0.18%
2010	100	1,024	670	0.19%
2011	54	1,077	336	0.19%
2012	216	1,293	1,251	0.21%
2013	275	1,568	1,466	0.25%
2014	326	1,895	1,592	0.28%
2015	70	1,965	309	0.28%

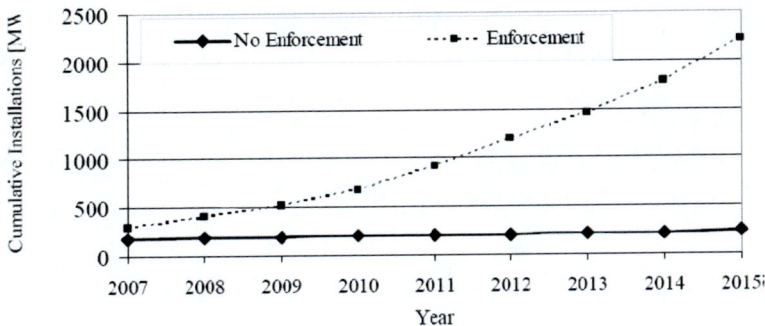

Figure 13. Impact of RPS solar set-asides, with all other scenarios at worst case.

4.2. The Base Case

The next case used more probable scenario inputs. An extension to the federal tax credits was assumed to pass (only to 2010 in the case of the residential tax credit, electricity prices were assumed to increase over time, carbon legislation was assumed to be enacted, and RPS solar set-asides were enforced, as detailed in table 8. We ran this scenario with BAU and SAI pricing to show not only the impact of the Solar America Initiative, but also what would happen if demand outpaced supply and prices do not decrease.

The positive impact on market penetration is noticeable compared with the worst case, as shown in the figures. The extension of the tax credits and RPS enforcement have the greatest impact. However, the market stalls temporarily in 2011 because the residential tax credit has expired. BAU system pricing yields a 26% compound annual growth rate (CAGR) to 2015. SAT system pricing results in a ~65% increase in cumulative installations over BAU pricing, with a 34%/year CAGR. State-by-state results are shown in the appendix.

Table 8. Base-Case Scenario Inputs

Scenario	Value
System Pricing Scenario	BAU/SAI
Interconnection Policy Scenario	Current Rules
Net Metering Availability Scenario	Current Availability
Net Metering Cap Scenario	Current Caps
Cap and Trade Scenario	Low Carbon Economy Act
Electricity Price Escalation	Medium
Federal Tax Credit	Extended
Time-of-Use Rates	Current Availability
RPS Solar Set Aside Enforcement	Yes

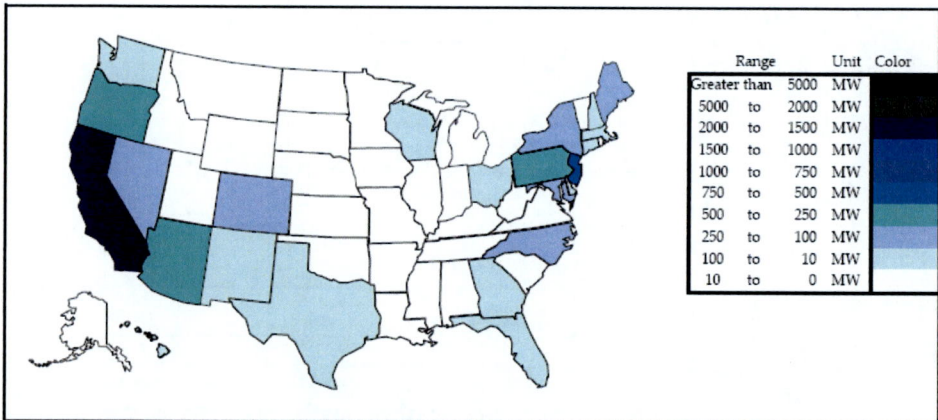

Figure 14. Cumulative installations in 2015 under the base case, with BAU system pricing.

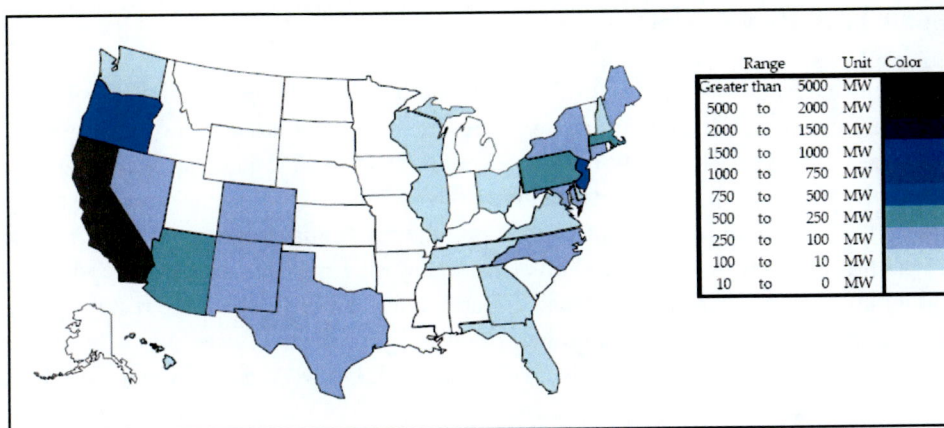

Figure 15. Cumulative installations in 2015 under the base case, with SAI system pricing.

Table 9. Nationwide Results for the Base Case, with BAU System Pricing

Year	Annual Installations [MW]	Cumulative Installation [MW]	Installers Required [FTE]	Market Penetration [%]
2007	251	733	1,864	0.16%
2008	833	1,567	4,885	0.33%
2009	223	1,790	1,554	0.35%
2010	288	2,078	1,937	0.39%
2011	270	2,348	1,687	0.41%
2012	527	2,875	3,055	0.48%
2013	313	3,188	1,668	0.50%
2014	544	3,732	2,654	0.55%
2015	813	4,545	3,588	0.64%

Table 10. Nationwide Results for the Base Case, with SAI System Pricing

Year	Annual Installations [MW]	Cumulative Installation [MW]	Installers Required [FTE]	Market Penetration [%]
2007	251	733	1,864	0.16%
2008	1,012	1,745	6,172	0.37%
2009	196	1,941	1,362	0.38%
2010	408	2,349	2,737	0.44%
2011	364	2,713	2,280	0.48%
2012	648	3,361	3,778	0.56%
2013	842	4,203	4,491	0.66%
2014	1,922	6,125	9,394	0.91%
2015	1,367	7,492	6,035	1.05%

4.3. Focused Policy Cases

Realizing that large amounts of effort are required to change state-level policies on a national scale, we took the two policies with the greatest impact and ran them together with the base case. Our analysis (shown in the appendix) found that improved net metering policy had the greatest impact on cumulative installations in 2015 (a 58% increase over the base case with SAI pricing). Next, fully extending the residential Investment Tax Credit (ITC) to 2015 had a 40% impact on cumulative installations. Table 11 shows the corresponding scenario inputs for the focused policy case. Figure 16 and table 12 show the results. With SAI system pricing, these two policies combine to increase cumulative installations by more than double by 2015 over the base-case, from 7,492 MW to 17,353 MW. State-by-state results can be found in the appendix.

Table 11. Focused Policy Case Inputs

Scenario	Value
System Pricing Scenario	BAU/SAI
Interconnection Policy Scenario	Current Rules
Net Metering Availability Scenario	Nationwide Availability
Net Metering Cap Scenario	Caps Lifted
Cap and Trade Scenario	Low Carbon Economy Act
Electricity Price Escalation	Accelerated
Federal Tax Credit	Fully Extended
Time-of-Use Rates	Current Availability
RPS Solar Set Aside Enforcement	Yes

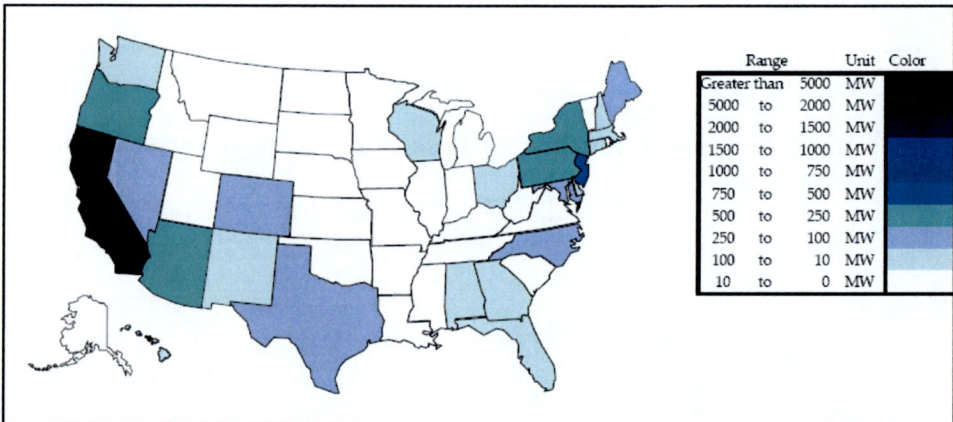

Figure 16. Cumulative installations in 2015 in the focused policy case, BAU system pricing.

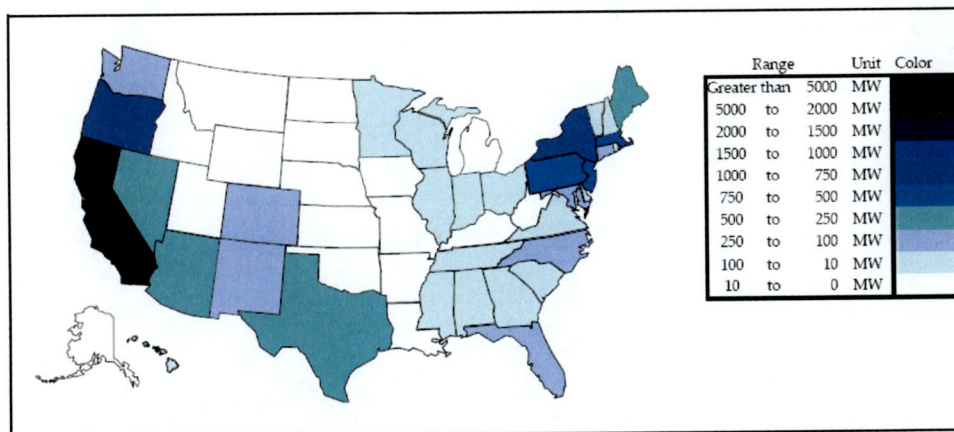

Figure 17. Cumulative installations in 2015 in the focused policy case, SAI system pricing.

Table 12. Nationwide Results for the Focused Policy Case, BAU System Pricing

Year	Annual Installations [MW]	Cumulative Installation [MW]	Installers Required [FTE]	Market Penetration [%]
2007	251	733	1,864	0.16%
2008	835	1,568	4,897	0.33%
2009	223	1,792	1,554	0.35%
2010	288	2,080	1,937	0.39%
2011	781	2,861	4,888	0.50%
2012	1,144	4,005	6,629	0.66%
2013	709	4,715	3,785	0.74%
2014	2,289	7,004	11,176	1.04%
2015	1,637	8,641	7,229	1.21%

Table 13. Nationwide Results for the Focused Policy Case, SAI System Pricing

Year	Annual Installations [MW]	Cumulative Installation [MW]	Installers Required [FTE]	Market Penetration [%]
2007	251	733	1,864	0.16%
2008	1,014	1,747	6,187	0.37%
2009	417	2,165	2,903	0.43%
2010	739	2,903	4,960	0.54%
2011	1,372	4,275	8,582	0.75%
2012	1,822	6,097	10,582	1.01%
2013	2,052	8,149	10,947	1.28%
2014	4,368	12,517	21,320	1.86%
2015	4,836	17,353	21,351	2.44%

4.4. The Best Case

The final case used inputs most favorable for PV market penetration, as shown in table 14. Figure 18 and table 15 show the national results. Achieving policy improvements in all these areas would require a large effort and potentially a considerable amount of federal funding. However, if this were successful, a very large, sustained demand (55%/year CAGR to 2015 with SAI pricing) can be created. State-by-state results are shown in the appendix.

Table 14. Best-Case Scenario Inputs

Scenario	Value
System Pricing Scenario	BAU/SAI
Interconnection Policy Scenario	Improved
Year of Policy Implementation	2008
Net Metering Availability Scenario	Nationwide Availability
Net Metering Cap Scenario	Caps Lifted
Cap and Trade Scenario	Low Carbon Economy Act
Electricity Price Escalation	Accelerated
Federal Tax Credit	Fully Extended
Time-of-Use Rates	Nationwide Availability
RPS Solar Set Aside Enforcement	Yes

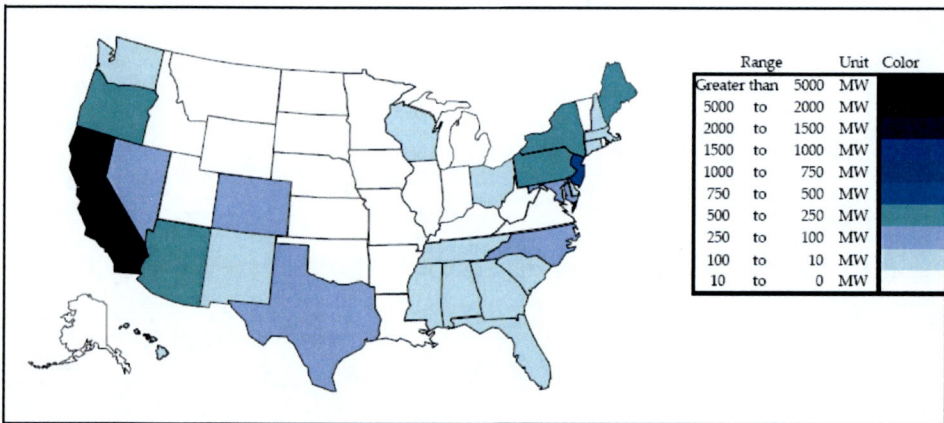

Figure 18. Cumulative installations in 2015 in the best case, BAU system pricing.

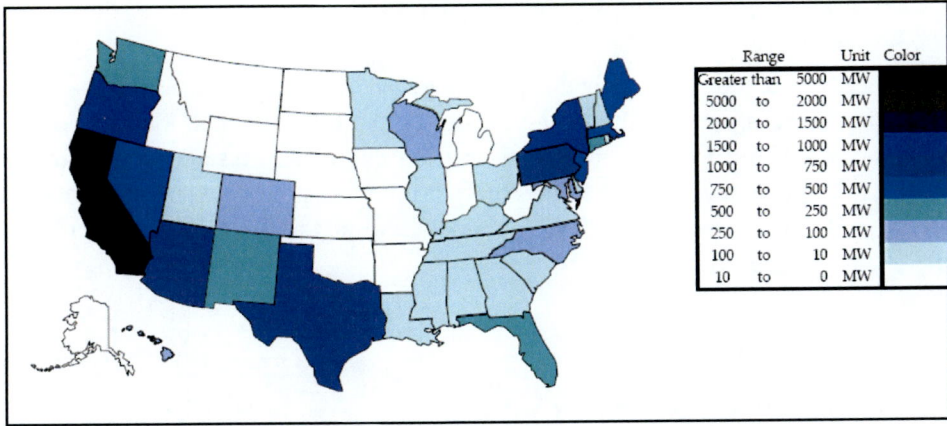

Figure 19. Cumulative installations in 2015 in the best case, SAI system pricing.

Table 15. Nationwide Results for the Best Case, BAU System Pricing

Year	Annual Installations [MW]	Cumulative Installation [MW]	Cumulative Installation [MW]	Market Penetration [%]
2007	251	733	1,864	0.16%
2008	1,019	1,753	6,226	0.37%
2009	314	2,067	2,183	0.41%
2010	420	2,487	2,822	0.46%
2011	1,004	3,491	6,282	0.61%
2012	1,372	4,864	7,953	0.81%
2013	1,045	5,909	5,577	0.93%
2014	2,633	8,542	12,886	1.27%
2015	2,565	11,107	11,326	1.56%

Table 16. Nationwide Results for the Best Case, SAI System Pricing

Year	Annual Installations [MW]	Cumulative Installation [MW]	Installers Required [FTE]	Market Penetration [%]
2007	251	733	1,864	0.16%
2008	1,237	1,970	7,793	0.41%
2009	622	2,593	4,328	0.51%
2010	1,187	3,780	7,974	0.70%
2011	1,496	5,276	9,357	0.92%
2012	2,383	7,659	13,868	1.27%
2013	2,807	10,466	14,989	1.64%
2014	6,724	17,190	32,780	2.55%
2015	7,522	24,712	33,208	3.47%

CONCLUSION

The critically important findings in this chapter are the influences of each scenario discussed. System pricing is the input with the largest impact. In the base case, the focused policy case, and the best case, using SAI system pricing caused cumulative installations to more than double by 2015. Other high-impact factors are net metering policy, extension of the federal tax credits, and interconnection policy. Figure 20 shows the cumulative effects of these variables.

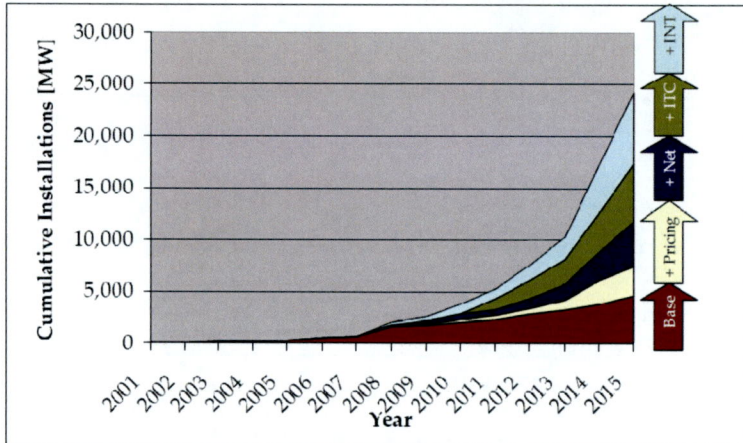

Figure 20. Influence of system pricing, net metering policy, federal tax credits, and interconnection policy on cumulative installations.

To understand the implication of these scenarios relative to planned generating capacity additions, we used data from EIA's 2007 Annual Energy Outlook. We compared the planned capacity projections in EIA's reference case from 2007 to 2015 to the cumulative installations of PV by 2015, as shown in table 17. Table 17 shows that PV could contribute between 27% to 91% of planned capacity additions per EIA's projections. Given that the U.S. market has strong regional variations, PV's contribution to capacity additions could be much higher on a regional or interconnect basis. This would have significant implications for utility planning and grid operations.

Table 17. Comparison of Planned Capacity Additions to Cumulative PV Installations with SAI Pricing

Scenario	EIA Projected Capacity Additions, 2007 to 2015 [MW]	2015 Cumulative PV Installations [MW]	PV as % of Planned Capacity Additions [%]
Base-Case	27,038	7,423	27%
Focused Policy Initiatives	27,038	17,353	64%
Best-Case	27,038	24,712	91%

During the course of this project, we identified several items that might enhance this analysis. The first would be an easily accessible database for building load profiles that might be similar to PV Watts for output profiles. Fortunately, NREL's commercial building load profiles were readily available for use, but the time required to generate profiles prevented us from using a unique residential profile for each utility analyzed. If a database of sample profiles were available, we could have used them for each utility's residential analysis.

Our analysis focused on rooftop applications, but other potential structures, such as parking garages or carports, are also suitable for PV installations. A useful activity might be to assess the feasibility of conducting a market potential analysis for PV on unoccupied structures. In addition, this study did not assess the potential for ground-mounted structures. A feasibility study should be conducted to identify or create methods and models for calculating the market potential for ground-mounted systems.

As discussed in Section 0, many groups within the PV industry, and those who monitor the PV industry (such as the investment community), have concerns about the availability of installers to meet a growing demand. For this study, we estimated installer requirements on a state-by-state basis for each year. However, it would provide valuable insights to model actual installer availability dynamics and feed the results back into the model.

The model we developed looks solely at the U.S. market and uses pricing assumptions that do not take into account demand outside the United States. If international markets (such as Spain or South Korea) experience dramatic surges in demand, module supplies could be diverted to those markets. A supply-constrained environment would then develop in the United States, however, and prices might not fall.

One key variable that the model does not now address is the impact of system financing. The market penetration curves used simple paybacks as inputs and did not consider financing. In reality, interest payments for financed systems affect economic attractiveness. Also, this model cannot assess the impact of innovative financing mechanisms or new business models (such as the power purchase agreement model) developing in the U.S. market. These drawbacks point to the need to develop a market penetration model based on return on investment or demand elasticity.

Finally, the model did not take into account possible electricity price feedbacks if the demand for grid power drops because of significant PV deployment.

However, even with these few shortcomings, this model reasonably simulates a very complex, intricate market by analyzing a large number of variables including system prices, electricity price forecasts, public policy, consumer behavior, and technology diffusion. The key findings of this study indicate that the technical potential and market opportunity for photovoltaics in the United States is significant if the government supports the appropriate policy mechanisms analyzed in the study.

REFERENCSE

California Energy Commission. (September 2007). *California Solar Initiative Program Handbook.* Available online at www.gosolarcalifornia.ca.gov.

Energy Information Administration. (2007). *Annual Energy Outlook 2007.* Available online at www.eia.doe.gov.

Energy Information Administration. (2003). *Commercial Building Energy Consumption Survey.* Available online at www.eia.doe.gov.

Energy Information Administration. *Form 861.* Available online at www.eia.doe.gov. Energy Information Administration. *Form 860.* Available online at www.eia.doe.gov.

Energy Information Administration. (2001). *Residential Energy Consumption Survey.* Available online at www.eia.doe.gov.

Fisher, J.C.; Pry, R.H. (1971). "A Simple Substitution Model of Technological Change." *Technological Forecasting and Social Change,* Vol. 3, pp. 75-88.

Network for New Energy Choices. (2007). *Freeing the Grid, 2007 Edition.* Available online at www.newenergychoices.org/uploads/FreeingTheGrid2007_report.pdf.

North Carolina Solar Center. (2007). Database of State Incentives for Renewables and Efficiency. Raleigh, NC: North Carolina State University; www.dsireusa.org.

Kastovich, J.C.; Lawrence, R.R.; Hoffman, R.R.; Pavlak, C. (1982). *Advanced Electric Heat Pump Market and Business Analysis.* Report no. ORNL/Sub/79-24712/1, prepared under subcontract to ORNL by Westinghouse Electric Corp. Oak Ridge, TN: Oak Ridge National Laboratory.

Roth, K.W.; Westphalen, D.; Dieckmann, J.; Hamilton, S.D.; Goetzler, W. (July 2002). *Energy Consumption Characteristics of Commercial Building HVAC Systems, Volume III: Energy Savings Potential.* Prepared by TIAX LLC for the DOE Building Technologies Program. Washington, DC: U.S. Department of Energy.

United States Senate Committee on Energy and Natural Resources. (July 2007). *Bingaman/Specter Climate Change Bill.* For information, see http://energy.senate.gov/public/.

APPENDIX: DETAILED RESULTS

A-1. Net Metering Improvements

After establishing a base case, NCI looked at the impact of lifting net-metering caps and allowing net metering in all states, as shown in table A-1. Figure A-1 and table A-2 show the cumulative installations in 2015 and nationwide results, respectively.

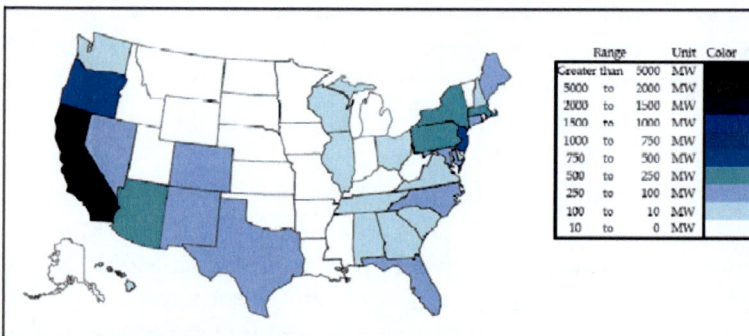

Figure A-1. Cumulative installations in 2015 in the net metering improvement case.

Table A-1. Net Metering Improvements
(Case Scenario Inputs)

Scenario	Value
System Pricing Scenario	SAI
Interconnection Policy Scenario	Current Rules
Net Metering Availability Scenario	Nationwide Availability
Net Metering Cap Scenario	Caps Lifted
Cap and Trade Scenario	Low Carbon Economy Act
Electricity Price Escalation	Accelerated
Federal Tax Credit	Extended
Time-of-Use Rates	Current Availability
RPS Solar Set Aside Enforcement	Yes

Table A-2. Nationwide Results for the Net Metering Improvement Case

Year	Annual Installations [MW]	Cumulative Installation [MW]	Installers Required [FTE]	Market Penetration [%]
2007	251	733	1,864	0.16%
2008	1,014	1,747	6,187	0.37%
2009	417	2,165	2,903	0.43%
2010	739	2,903	4,960	0.54%
2011	329	3,232	2,059	0.57%
2012	1,140	4,372	6,630	0.72%
2013	1,333	5,705	7,109	0.89%
2014	3,136	8,841	15,311	1.31%
2015	2,973	11,813	13,124	1.66%

Lifting net metering caps and establishing net metering have noticeable impacts in a few states—California, Florida, New York, and Oregon. This means that installations do not reach net-metering cap amounts in any other states, and net metering improves system economics in states that do not allow net metering. California has a net-metering cap of 2.5% of a utility's peak load, New York has a net metering cap of 0.1% of a utility's peak load, and Oregon has a net metering cap of 0.5% of a utility's peak load. Florida does not currently allow net metering. Figure A-2 shows the combined impact of improved net-metering policies in these states, but most is driven by California.

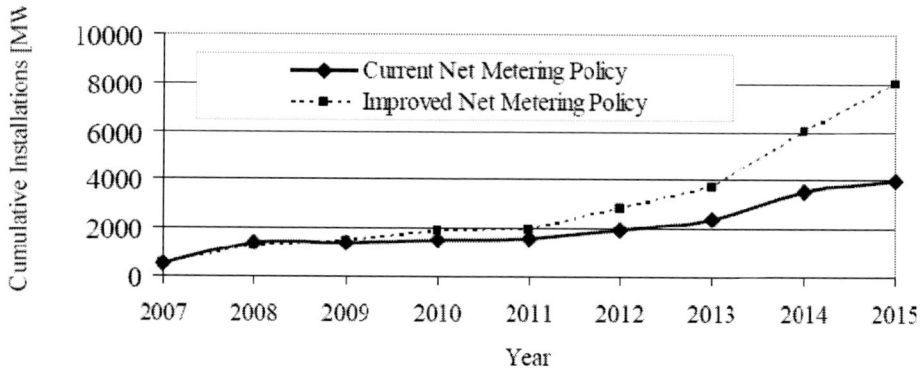

Figure A-2. Impact of improved net metering policies in California, Florida, New York, and Oregon.

A-2. Interconnection Standard Improvements

The next case started back at the base case and looked at improved interconnection standards, as shown in table A-3. Many states (or utilities) have interconnection standards that inhibit PV adoption. However, many state legislatures are in the process of revising their interconnection standards. This case examines the impact of all states improving their interconnection standards to "superior" per the IREC rating shown in table 2 and assumes that improved standards are in place by 2008. Results are shown in Figure A-3 and table A-4.

Table A-3. Interconnection Standard Improvements Case Scenario Inputs

Scenario	Value
System Pricing Scenario	SAI
Interconnection Policy Scenario	Improved
Year of Policy Implementation	2008
Net Metering Availability Scenario	Current Availability
Net Metering Cap Scenario	Business-As-Usual
Cap and Trade Scenario	Low Carbon Economy Act
Electricity Price Escalation	Accelerated
Federal Tax Credit	Extended
Time-of-Use Rates	Current Availability
RPS Solar Set-Aside Enforcement	Yes

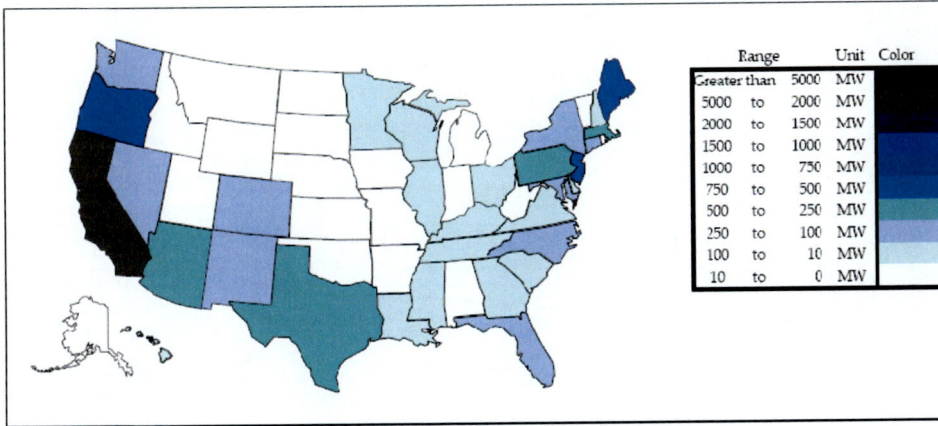

Figure A-3. Cumulative installations in 2015 in the interconnection standards improvement case.

Table A-4. Nationwide Results for the Interconnection Standards Improvement Case

Year	Annual Installations [MW]	Cumulative Installation [MW]	Installers Required [FTE]	Market Penetration [%]
2007	251	733	1,864	0.16%
2008	1,221	1,955	7,678	0.41%
2009	284	2,239	1,979	0.44%
2010	797	3,036	5,350	0.56%
2011	300	3,336	1,876	0.58%
2012	948	4,284	5,494	0.71%
2013	821	5,104	4,399	0.80%
2014	2,603	7,707	12,731	1.14%
2015	1,899	9,606	8,398	1.35%

Improving interconnection standards has a large impact in the following states, which have interconnection assessments of "poor" or below: Connecticut (poor), Florida (poor), Hawaii (barrier), Illinois (barrier), Maine (barrier), Pennsylvania (poor), Washington (barrier), and Wisconsin (poor). Figure A-4 shows a combined increase of ~60% in cumulative installations by 2015 in these states if interconnection standards are improved.

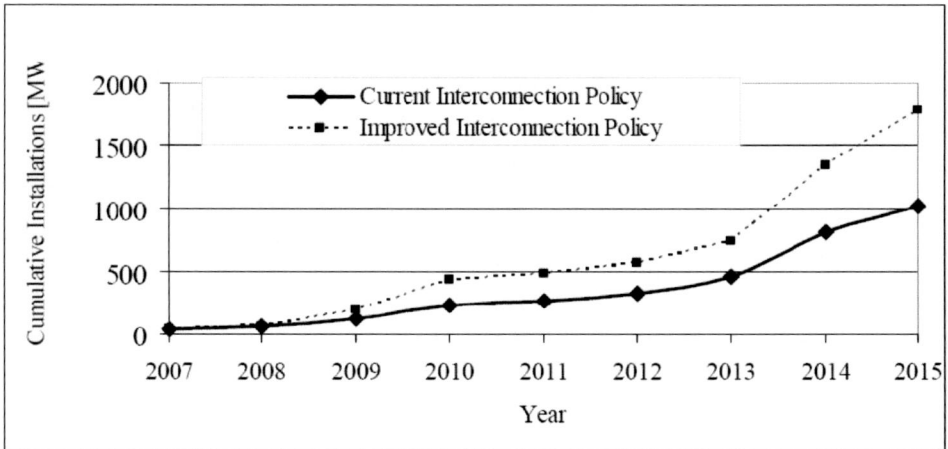

Figure A-4. Result of improved interconnection standards in Connecticut, Florida, Hawaii, Illinois, Maine, Pennsylvania, Washington, and Wisconsin.

A-3. Nationwide Availability of Time-of-Use Rates

The next case run assumed that TOU rates were available from every utility, as shown in table A-5. We reviewed the economics in each utility region to determine if standard or TOU rates resulted in lower annual electric bills and then chose the cheaper option. This yielded some interesting results (see figure A-5 and table A-6). Some utilities in Hawaii (specifically, Maui Electric Company) and Texas (all the utilities analyzed except Entergy Gulf States) do not have TOU rates, so this increased penetration. However, the establishment of TOU rates actually decreases market penetration in Massachusetts, New Jersey, and Tennessee. Some utilities in these states do not offer TOU rates; implementing them results in lower electric bills, which in turn results in lower annual electric bill savings as a result of using PV. Thus, the simple payback increases and market penetration decreases.

Table A-5. Time-of-Use Availability Scenario Inputs

Scenario	Value
System Pricing Scenario	SAI
Interconnection Policy Scenario	Current Rules
Net Metering Availability Scenario	Current Availability
Net Metering Cap Scenario	Business-As-Usual
Cap and Trade Scenario	Low Carbon Economy Act
Electricity Price Escalation	Accelerated
Federal Tax Credit	Extended
Time-of-Use Rates	Nationwide Availability
RPS Solar Set Aside Enforcement	Yes

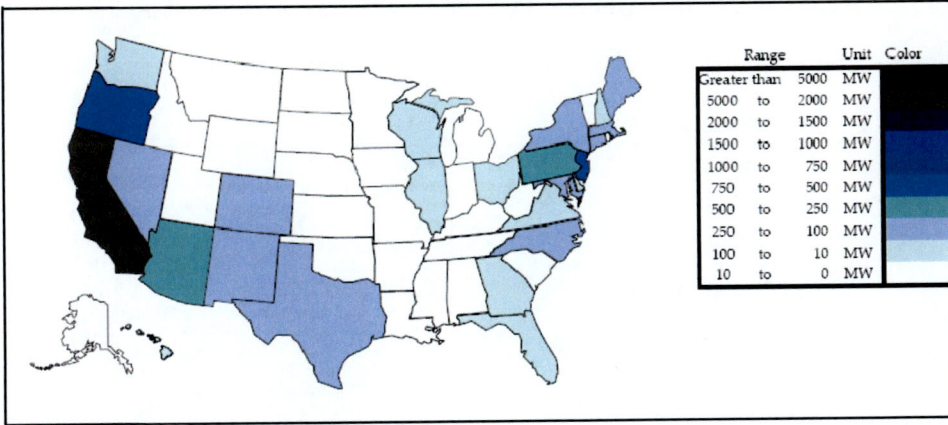

Figure A-5. Cumulative installations in 2015 in the time-of-use availability case.

Table A-6. Nationwide Results for the Time-of-Use Availability Case

Year	Annual Installations [MW]	Cumulative Installation [MW]	Installers Required [FTE]	Market Penetration [%]
2007	251	733	1,864	0.16%
2008	1,016	1,749	6,200	0.37%
2009	201	1,950	1,396	0.38%
2010	411	2,361	2,762	0.44%
2011	360	2,722	2,254	0.48%
2012	638	3,359	3,720	0.56%
2013	841	4,201	4,488	0.66%
2014	1,845	6,046	9,019	0.90%
2015	1,370	7,415	6,048	1.04%

A-4. Fully Extended Residential Federal Tax Credit

To look at the impact of the federal tax credit, we assumed the residential federal tax credit would be extended until 2016. Table A-7 shows the scenario inputs, while figure A-6 and table A-8 show the resulting cumulative installations. The extension affects all markets, but the impacts are strongest in California, Massachusetts, Pennsylvania, and Texas, as shown in figure A-7.

Table A-7. Fully Extended Federal Tax Credit Scenario Inputs

Scenario	Value
System Pricing Scenario	SAI
Interconnection Policy Scenario	Current Rules
Net Metering Availability Scenario	Current Availability
Net Metering Cap Scenario	Business-As-Usual
Cap and Trade Scenario	Low Carbon Economy Act
Electricity Price Escalation	Accelerated
Federal Tax Credit	Fully Extended
Time-of-Use Rates	Current Availability
RPS Solar Set Aside Enforcement	Yes

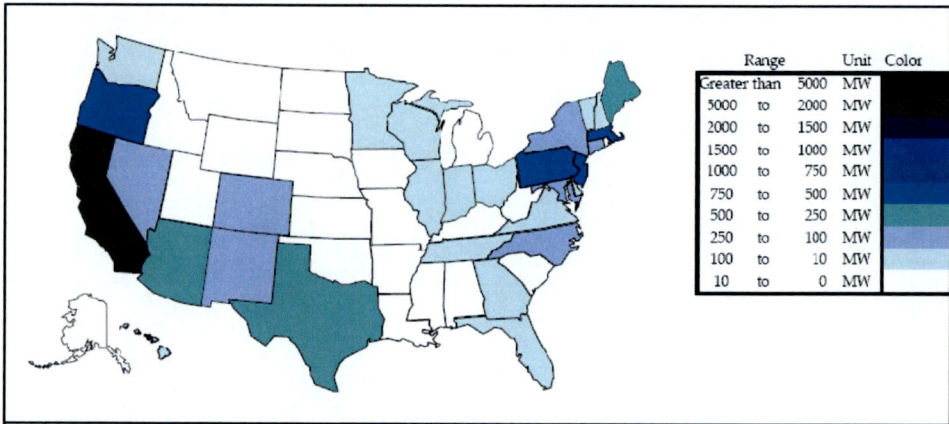

Figure A-6. Cumulative installations in 2015: fully extended tax credit case.

Table A-8. Nationwide Results for the Fully Extended Tax Credit Case

Year	Annual Installations [MW]	Cumulative Installation [MW]	Installers Required [FTE]	Market Penetration [%]
2007	251	733	1,864	0.16%
2008	1,012	1,745	6,172	0.37%
2009	196	1,941	1,362	0.38%
2010	408	2,349	2,737	0.44%
2011	562	2,911	3,520	0.51%
2012	1,097	4,008	6,378	0.66%
2013	655	4,663	3,497	0.73%
2014	2,292	6,955	11,196	1.03%
2015	3,044	9,998	13,438	1.40%

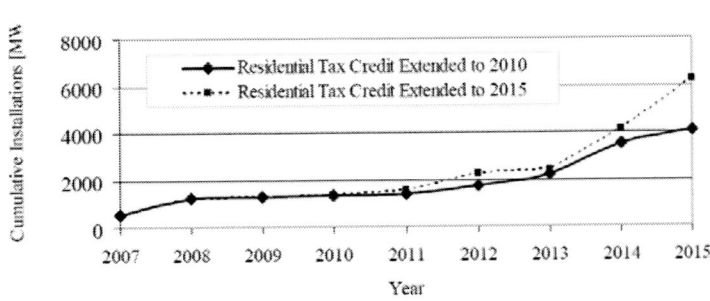

Figure A-7. Impact of extending the residential federal tax credit through 2015 in California, Connecticut, Pennsylvania, and Texas.

A-5. State-by-State Results

Table A-9. State-by-State Technical Potential, Over Time

	Alabama	Alaska	Arizona	Arkansas	California	Colorado	Connecticut	Delaware	Florida	Georgia	Hawaii	Idaho	Illinois
2007	9,376	840	10,515	4,655	51,667	7,778	3,986	1,217	34,087	16,574	1,883	2,194	17,594
2008	9,989	889	11,455	4,948	54,975	8,350	4,197	1,312	37,062	17,915	2,003	2,379	18,604
2009	10,601	943	12,447	5,261	58,344	8,955	4,414	1,400	40,062	19,321	2,119	2,572	19,648
2010	11,227	997	13,499	5,552	61,835	9,596	4,636	1,489	43,070	20,771	2,242	2,770	20,705
2011	11,855	1,050	14,579	5,849	65,377	10,249	4,858	1,579	46,133	22,254	2,366	2,968	21,771
2012	12,495	1,104	15,701	6,153	69,021	10,923	5,087	1,674	49,394	23,802	2,493	3,172	22,848
2013	13,178	1,161	16,946	6,479	72,828	11,644	5,317	1,777	52,985	25,487	2,626	3,402	23,974
2014	13,882	1,219	18,268	6,815	76,753	12,397	5,552	1,885	56,770	27,257	2,762	3,645	25,125
2015	14,606	1,279	19,671	7,160	80,798	13,184	5,790	1,997	60,760	29,119	2,903	3,901	26,302

	Indiana	Iowa	Kansas	Kentucky	Louisiana	Maine	Maryland	Mass.	Michigan	Minnesota	Mississippi	Missouri
2007	9,909	4,602	4,444	7,596	8,359	1,483	8,203	6,959	14,347	8,081	5,207	8,487
2008	10,521	4,867	4,700	8,065	8,887	1,569	8,730	7,329	15,137	8,571	5,534	9,014
2009	11,167	5,140	4,968	8,562	9,431	1,654	9,262	7,704	15,958	9,087	5,860	9,549
2010	11,822	5,418	5,242	9,068	9,954	1,742	9,804	8,091	16,792	9,609	6,198	10,092
2011	12,487	5,692	5,521	9,575	10,484	1,831	10,356	8,482	17,635	10,137	6,537	10,639
2012	13,167	5,970	5,805	10,089	11,022	1,923	10,921	8,882	18,500	10,683	6,876	11,201
2013	13,881	6,264	6,098	10,635	11,599	2,017	11,517	9,285	19,385	11,248	7,240	11,790
2014	14,617	6,564	6,397	11,198	12,192	2,113	12,130	9,695	20,288	11,829	7,615	12,395
2015	15,376	6,870	6,703	11,777	12,800	2,212	12,761	10,111	21,211	12,427	7,999	13,016

	Montana	Nebraska	Nevada	NH	New Jersey	New Mexico	New York	North Carolina	North Dakota	Ohio	Oklahoma	Oregon	Pennsylvania
2007	1,234	2,712	5,040	1,413	7,801	2,852	14,521	15,144	1,040	18,159	6,399	5,231	11,362
2008	1,304	2,881	5,535	1,499	8,228	3,036	15,262	16,234	1,099	19,182	6,775	5,581	11,969
2009	1,376	3,051	6,061	1,588	8,685	3,230	16,011	17,398	1,159	20,208	7,169	5,962	12,605
2010	1,450	3,226	6,615	1,679	9,138	3,437	16,766	18,569	1,219	21,266	7,561	6,351	13,246
2011	1,525	3,402	7,177	1,771	9,596	3,645	17,520	19,762	1,279	22,331	7,963	6,747	13,886
2012	1,601	3,583	7,760	1,866	10,064	3,858	18,285	21,010	1,339	23,420	8,370	7,152	14,539
2013	1,680	3,772	8,429	1,965	10,546	4,080	19,067	22,341	1,403	24,540	8,795	7,582	15,209
2014	1,760	3,967	9,145	2,066	11,036	4,311	19,858	23,728	1,468	25,682	9,230	8,029	15,891
2015	1,842	4,167	9,911	2,170	11,536	4,549	20,659	25,175	1,534	26,847	9,675	8,492	16,585

Table A-9 (Continued)

	Rhode Island	South Carolina	South Dakota	Tennessee	Texas	Utah	Vermont	Virginia	Washington	DC	West Virginia	Wisconsin	Wyoming
2007	1,036	7,619	1,106	11,774	42,773	3,691	708	13,565	9,025	1,236	2,467	8,158	768
2008	1,090	8,208	1,174	12,561	45,863	3,985	749	14,506	9,646	1,297	2,599	8,647	816
2009	1,145	8,817	1,245	13,370	49,089	4,279	789	15,444	10,309	1,369	2,728	9,139	865
2010	1,200	9,422	1,317	14,206	52,320	1,603	830	16,421	10,989	1,443	2,858	9,649	914
2011	1,255	10,039	1,388	15,049	55,632	4,927	872	17,417	11,681	1,516	2,905	10,165	964
2012	1,312	10,694	1,461	15,912	59,039	5,261	915	18,448	12,395	1,588	3,112	10,696	1,013
2013	1,369	11,398	1,538	16,829	62,708	5,625	959	19,538	13,152	1,663	3,246	11,246	1,067
2014	1,428	12,133	1,618	17,778	66,527	6,006	1,004	20,667	13,338	1,740	3,383	11,810	1,123
2015	1,487	12,902	1,700	18,757	70,499	6,407	1,050	21,837	14,755	1,818	3,522	12,389	1,180

Table A-10. State-by-State Results for the Worst Case

		Alabama	Alaska	Arizona	Arkansas	California	Colorado	Connecticut	Delaware	Florida	Georgia	Hawaii	Idaho	Illinois
Cumulative Installations	2007	1	1	14	1	499	20	3	1	1	1	6	1	1
	2008	1	1	21	1	598	23	3	5	2	1	9	1	1
	2009	1	1	21	1	625	23	3	5	3	1	9	1	1
	2010	1	1	21	1	678	23	4	5	4	1	9	1	1
	2011	1	1	21	1	723	23	4	5	4	1	9	1	1
	2012	1	1	21	1	923	23	5	5	4	1	9	1	1
	2013	1	1	21	1	1,164	25	8	5	4	1	9	1	1
	2014	1	1	21	1	1,445	27	10	5	4	1	9	1	1
	2015	1	1	21	1	1,445	30	13	5	4	1	9	1	1
Annual Installations	2007	0	0	5	0	166	6	1	1	1	0	3	0	0
	2008	0	0	7	0	100	2	0	4	1	0	3	0	0
	2009	0	0	0	0	27	0	0	0	1	0	0	0	0
	2010	0	0	0	0	53	0	0	0	1	0	0	0	0
	2011	0	0	0	0	45	0	1	0	0	0	0	0	0
	2012	0	0	0	0	199	1	1	0	0	0	0	0	0
	2013	0	0	0	0	241	2	2	0	0	0	0	0	0
	2014	0	0	0	0	281	2	2	0	0	0	0	0	0
	2015	0	0	0	0	0	3	3	0	0	0	0	0	0
Installers Required	2007	0	0	35	0	1,232	48	7	4	4	0	23	0	0
	2008	0	0	52	0	717	18	2	26	6	0	23	1	0
	2009	0	0	0	0	189	0	2	0	6	0	0	0	0
	2010	0	0	0	0	357	0	2	0	6	0	0	0	0
	2011	0	0	0		281	0	4	0	0	0	0	0	0
	2012	0	0	0	0	1,155	4	8	0	0	0	0	0	0
	2013	0	0	0	0	1,286	9	11	0	0	0	0	0	0
	2014	0	0	0	0	1,369	8	10	0	0	0	0	0	0
	2015	0	0	0	0	0	13	15	0	0	0	0	0	0
Market Penetration	2007	0%	0%	0%	0%	1%	0%	0%	0%	0%	0%	0%	0%	0%
	2008	0%	0%	0%	0%	1%	0%	0%	0%	0%	0%	0%	0%	0%
	2009	0%	0%	0%	0%	1%	0%	0%	0%	0%	0%	0%	0%	0%
	2010	0%	0%	0%	0%	1%	0%	0%	0%	0%	0%	0%	0%	0%
	2011	0%	0%	0%	0%	1%	0%	0%	0%	0%	0%	0%	0%	0%
	2012	0%	0%	0%	0%	1%	0%	0%	0%	0%	0%	0%	0%	0%
	2013	0%	0%	0%	0%	2%	0%	0%	0%	0%	0%	0%	0%	0%
	2014	0%	0%	0%	0%	2%	0%	0%	0%	0%	0%	0%	0%	0%
	2015	0%	0%	0%	0%	2%	0%	0%	0%	0%	0%	0%	0%	0%

		Indiana	Iowa	Kansas	Kentucky	Louisiana	Maine	Maryland	Mass.	Michigan	Minnesota	Mississippi	Missouri
Cumulative Installations	2007	1	1	1	1	1	2	2	7	1	1	1	1
	2008	1	1	1	1	1	2	2	9	1	1	1	1
	2009	1	1	1	1	1	5	2	10	1	1	1	1
	2010	1	1	1	1	1	42	2	12	1	1	1	1
	2011	1	1	1	1	1	42	2	12	1	1	1	1
	2012	1	1	1	1	1	42	2	14	1	1	1	1
	2013	1	1	1	1	1	51	2	18	1	1	1	1
	2014	1	1	1	1	1	67	2	23	1	1	1	1
	2015	1	1	1	1	1	90	3	31	1	1	1	1
Annual Installations	2007	0	0	0	0	0	1	1	2	1	0	0	0
	2008	0	0	0	1	0	0	0	2	0	0	0	0
	2009	0	0	0	0	0	2	0	2	0	0	0	0
	2010	0	0	0	0	0	37	0	2	0	0	0	0
	2011	0	0	0	0	0	0	0	0	0	0	0	0
	2012	0	0	0	0	0	0	0	2	0	0	0	0
	2013	0	0	0	0	0	9	0	5	0	0	0	0
	2014	0	0	0	0	0	16	0	5	0	0	0	0
	2015	0	0	0	0	0	23	1	7	0	0	0	0
Installers Required	2007	0	0	0	0	0	7	7	16	4	0	0	0
	2008	0	0	0	7	0	1	0	13	0	2	0	0
	2009	0	0	0	0	0	17	0	13	0	0	0	0
	2010	0	0	0	0	0	250	0	12	0	0	0	0
	2011	0	0	0	0	0	0	0	0	0	0	0	0
	2012	0	0	0	0	0	0	0	9	0	0	0	0
	2013	0	0	0	0	0	47	0	25	0	0	0	0
	2014	0	0	0	0	0	80	0	22	0	0	0	0
	2015	0	0	0	0	0	100	3	33	0	0	0	0
Market Penetration	2007	0%	0%	0%	0%	0%	0%	0%	0%	0%	0%	0%	0%
	2008	0%	0%	0%	0%	0%	0%	0%	0%	0%	0%	0%	0%
	2009	0%	0%	0%	0%	0%	0%	0%	0%	0%	0%	0%	0%
	2010	0%	0%	0%	0%	0%	2%	0%	0%	0%	0%	0%	0%
	2011	0%	0%	0%	0%	0%	2%	0%	0%	0%	0%	0%	0%
	2012	0%	0%	0%	0%	0%	2%	0%	0%	0%	0%	0%	0%
	2013	0%	0%	0%	0%	0%	3%	0%	0%	0%	0%	0%	0%
	2014	0%	0%	0%	0%	0%	3%	0%	0%	0%	0%	0%	0%
	2015	0%	0%	0%	0%	0%	4%	0%	0%	0%	0%	0%	0%

Table A-10 (Continued)

		Montana	Nebraska	Nevada	NH	New Jersey	New Mexico	New York	North Carolina	North Dakota	Ohio	Oklahoma	Oregon	Pennsylvania
Cumulative Installations	2007	1	1	15	1	69	9	32	3	1	2	1	2	9
	2008	1	1	16	1	73	12	35	6	1	6	1	6	9
	2009	1	1	16	1	73	12	35	6	1	6	1	9	9
	2010	1	1	16	1	73	12	36	9	1	6	1	9	9
	2011	1	1	16	1	73	12	38	9	1	6	1	11	9
	2012	1	1	17	1	73	12	41	9	1	6	1	14	10
	2013	1	1	18	1	73	12	45	9	1	6	1	16	11
	2014	1	1	19	2	73	12	49	9	1	6	1	20	13
	2015	1	1	21	3	73	14	56	9	1	6	1	24	17
Annual Installations	2007	0	0	7	1	29	3	10	1	0	1	0	1	3
	2008	0	0	1	0	3	3	3	3	0	4	0	4	0
	2009	0	0	0	0	0	0	0	0	0	0	0	3	0
	2010	0	0	0	0	0	0	0	2	0	0	0	0	0
	2011	0	0	0	0	0	0	2	0	0	0	0	2	0
	2012	0	0	1	0	0	0	3	0	0	0	0	3	1
	2013	0	0	1	0	0	0	4	0	0	0	0	3	1
	2014	0	0	1	1	0	0	4	0	0	0	0	3	2
	2015	0	0	2	1	0	2	7	0	0	0	0	4	4
Installers Required	2007	0	0	56	4	213	24	71	7	0	4	0	4	19
	2008	3	0	6	0	22	21	20	24	0	26	0	30	0
	2009	0	0	0	0	0	0	0	0	0	0	0	18	0
	2010	0	0	0	0	0	0	3	16	0	0	0	0	0
	2011	0	0	2	0	0	0	14	0	0	0	0	14	0
	2012	0	0	3	0	0	0	19	0	0	0	0	15	3
	2013	0	0	5	2	0	0	20	0	0	0	0	15	7
	2014	0	0	6	3	0	2	18	0	0	0	0	17	9
	2015	0	0	8	5	0	7	30	0	0	0	0	19	16
Market Penetration	2007	0%	0%	0%	0%	1%	0%	0%	0%	0%	0%	0%	0%	0%
	2008	0%	0%	0%	0%	1%	0%	0%	0%	0%	0%	0%	0%	0%
	2009	0%	0%	0%	0%	1%	0%	0%	0%	0%	0%	0%	0%	0%
	2010	0%	0%	0%	0%	1%	0%	0%	0%	0%	0%	0%	0%	0%
	2011	0%	0%	0%	0%	1%	0%	0%	0%	0%	0%	0%	0%	0%
	2012	0%	0%	0%	0%	1%	0%	0%	0%	0%	0%	0%	0%	0%
	2013	0%	0%	0%	0%	1%	0%	0%	0%	0%	0%	0%	0%	0%
	2014	0%	0%	0%	0%	1%	0%	0%	0%	0%	0%	0%	0%	0%
	2015	0%	0%	0%	0%	1%	0%	0%	0%	0%	0%	0%	0%	0%

		Rhode Island	South Carolina	South Dakota	Tennessee	Texas	Utah	Vermont	Virginia	Washington	DC	West Virginia	Wisconsin	Wyoming
Cumulative Installations	2007	1	2	1	1	8	1	2	1	6	1	1	6	1
	2008	1	2	1	6	9	1	2	1	8	1	1	9	1
	2009	1	2	1	6	9	1	2	1	8	1	1	9	1
	2010	1	2	1	6	12	1	2	1	8	1	1	9	1
	2011	1	2	1	6	15	1	2	1	9	1	1	9	1
	2012	1	2	1	6	20	1	2	1	10	1	1	9	1
	2013	1	2	1	6	25	1	2	1	11	1	1	9	1
	2014	1	2	1	6	33	1	2	1	13	2	1	9	1
	2015	1	2	1	6	45	1	2	1	13	2	1	10	1
Annual Installations	2007	1	1	0	0	1	1	1	1	3	1	0	2	0
	2008	0	0	0	5	3	0	0	1	2	0	0	3	0
	2009	0	0	0	0	0	0	0	0	0	0	0	0	0
	2010	0	0	0	0	3	0	0	0	0	0	0	0	0
	2011	0	0	0	0	3	0	0	0	1	0	0	0	0
	2012	0	0	0	0	5	0	0	0	1	0	0	0	0
	2013	0	0	0	0	6	0	0	0	1	0	0	0	0
	2014	0	0	0	0	8	0	0	0	1	0	0	0	0
	2015	0	0	0	0	12	0	0	0	0	0	0	1	0
Installers Required	2007	4	7	0	0	8	4	7	4	22	4	0	15	0
	2008	0	0	0	36	20	0	1	4	13	0	0	22	0
	2009	0	0	0	0	0	0	0	0	0	0	0	0	0
	2010	0	0	0	0	22	0	0	0	2	1	0	0	0
	2011	0	0	0	0	17	0	0	0	5	0	0	0	0
	2012	0	0	0	0	29	0	0	0	6	1	0	0	0
	2013	0	0	0	0	31	0	0	0	7	1	0	0	0
	2014	0	0	0	0	37	0	0	0	7	1	0	0	0
	2015	1	0	0	0	52	0	1	0	0	1	0	3	0
Market Penetration	2007	0%	0%	0%	0%	0%	0%	0%	0%	0%	0%	0%	0%	0%
	2008	0%	0%	0%	0%	0%	0%	0%	0%	0%	0%	0%	0%	0%
	2009	0%	0%	0%	0%	0%	0%	0%	0%	0%	0%	0%	0%	0%
	2010	0%	0%	0%	0%	0%	0%	0%	0%	0%	0%	0%	0%	0%
	2011	0%	0%	0%	0%	0%	0%	0%	0%	0%	0%	0%	0%	0%
	2012	0%	0%	0%	0%	0%	0%	0%	0%	0%	0%	0%	0%	0%
	2013	0%	0%	0%	0%	0%	0%	0%	0%	0%	0%	0%	0%	0%
	2014	0%	0%	0%	0%	0%	0%	0%	0%	0%	0%	0%	0%	0%
	2015	0%	0%	0%	0%	0%	0%	0%	0%	0%	0%	0%	0%	0%

Table A-11. State-by-State Results for the Base-Case, with BAU System Pricing

		Alabama	Alaska	Arizona	Arkansas	California	Colorado	Connecticut	Delaware	Florida	Georgia	Hawaii	Idaho	Illinois
Cumulative Installations	2007	1	1	14	1	499	20	3	1	1	1	6	1	1
	2008	1	1	41	1	1,051	34	3	10	2	1	10	1	1
	2009	1	1	72	1	1,105	35	6	10	3	1	10	1	1
	2010	1	1	122	1	1,105	36	8	10	4	1	10	1	1
	2011	1	1	187	1	1,105	73	8	10	5	1	10	1	1
	2012	1	1	268	1	1,272	75	10	10	9	1	10	1	1
	2013	1	1	313	1	1,272	77	14	10	13	1	10	1	1
	2014	1	1	360	1	1,275	78	21	16	21	9	10	1	1
	2015	1	1	408	1	1,566	120	24	20	30	11	12	1	2
Annual Installations	2007	0	0	5	0	166	6	1	1	1	0	3	0	0
	2008	0	0	27	0	552	14	0	9	1	0	4	0	0
	2009	0	0	31	0	55	1	3	0	1	0	0	0	0
	2010	0	0	50	0	0	1	1	0	1	0	0	0	0
	2011	0	0	65	0	0	37	0	0	1	0	0	0	0
	2012	0	0	81	0	167	2	3	0	4	0	0	0	0
	2013	0	0	45	0	0	2	4	0	3	1	0	0	0
	2014	0	0	47	0	3	2	7	6	9	7	0	0	0
	2015	0	0	49	0	291	42	4	4	9	3	1	0	1
Installers Required	2007	0	0	35	0	1,232	48	7	4	4	0	23	0	0
	2008	0	0	171	0	3,973	102	2	64	6	0	27	1	0
	2009	0	0	214	0	379	5	24	0	6	0	0	1	0
	2010	0	0	339	0	0	5	9	0	6	0	0	0	0
	2011	0	0	407	0	0	235	0	0	8	0	0	0	0
	2012	0	0	467	0	965	9	16	0	26	0	0	0	0
	2013	0	0	239	0	0	9	19	0	17	5	0	0	0
	2014	0	0	228	0	15	8	33	28	44	35	2	0	0
	2015	0	0	214	0	1,286	184	16	18	38	12	6	0	5
Market Penetration	2007	0%	0%	0%	0%	1%	0%	0%	0%	0%	0%	0%	0%	0%
	2008	0%	0%	0%	0%	2%	0%	0%	1%	0%	0%	0%	0%	0%
	2009	0%	0%	1%	0%	2%	0%	0%	1%	0%	0%	0%	0%	0%
	2010	0%	0%	1%	0%	2%	0%	0%	1%	0%	0%	0%	0%	0%
	2011	0%	0%	1%	0%	2%	1%	0%	1%	0%	0%	0%	0%	0%
	2012	0%	0%	2%	0%	2%	1%	0%	1%	0%	0%	0%	0%	0%
	2013	0%	0%	2%	0%	2%	1%	0%	1%	0%	0%	0%	0%	0%
	2014	0%	0%	2%	0%	2%	1%	0%	1%	0%	0%	0%	0%	0%
	2015	0%	0%	2%	0%	2%	1%	0%	1%	0%	0%	0%	0%	0%

		Indiana	Iowa	Kansas	Kentucky	Louisiana	Maine	Maryland	Mass.	Michigan	Minnesota	Mississippi	Missouri
Cumulative Installations	2007	1	1	1	1	1	2	2	7	1	1	1	1
	2008	1	1	1	2	1	2	3	9	1	1	1	1
	2009	1	1	1	2	1	25	15	10	1	1	1	1
	2010	1	1	1	2	1	79	24	12	1	1	1	1
	2011	1	1	1	2	1	79	24	18	1	1	1	1
	2012	1	1	1	2	1	79	59	26	1	1	1	1
	2013	1	1	1	2	1	79	59	34	1	1	1	1
	2014	1	1	1	2	1	96	149	49	2	1	2	1
	2015	1	1	1	2	1	122	149	57	2	1	2	1
Annual Installations	2007	0	0	0	0	0	1	1	2	1	0	0	0
	2008	0	0	0	1	0	0	1	2	0	0	0	0
	2009	0	0	0	0	0	22	13	2	0	0	0	0
	2010	0	0	0	0	0	55	9	2	0	0	0	0
	2011	0	0	0	0	0	0	0	5	0	0	0	0
	2012	0	0	0	0	0	0	35	9	0	0	0	0
	2013	0	0	0	0	0	0	0	8	0	0	0	0
	2014	0	0	0	0	0	17	90	16	1	0	1	0
	2015	0	0	0	0	0	25	0	8	1	0	1	0
Installers Required	2007	0	0	0	0	0	7	7	16	4	0	0	0
	2008	0	0	0	8	0	3	4	13	0	2	0	0
	2009	0	0	0	0	0	155	88	13	0	0	0	0
	2010	0	0	0	0	0	367	57	12	0	0	0	0
	2011	0	0	0	0	0	0	0	33	0	0	0	0
	2012	0	0	0	0	0	0	202	49	0	0	2	0
	2013	0	0	0	0	0	0	0	41	0	0	1	0
	2014	0	0	0	0	0	84	441	76	5	0	3	0
	2015	0	0	0	0	0	111	0	33	3	0	3	0
Market Penetration	2007	0%	0%	0%	0%	0%	0%	0%	0%	0%	0%	0%	0%
	2008	0%	0%	0%	0%	0%	0%	0%	0%	0%	0%	0%	0%
	2009	0%	0%	0%	0%	0%	1%	0%	0%	0%	0%	0%	0%
	2010	0%	0%	0%	0%	0%	5%	0%	0%	0%	0%	0%	0%
	2011	0%	0%	0%	0%	0%	4%	0%	0%	0%	0%	0%	0%
	2012	0%	0%	0%	0%	0%	4%	1%	0%	0%	0%	0%	0%
	2013	0%	0%	0%	0%	0%	4%	1%	0%	0%	0%	0%	0%
	2014	0%	0%	0%	0%	0%	5%	1%	1%	0%	0%	0%	0%
	2015	0%	0%	0%	0%	0%	6%	1%	1%	0%	0%	0%	0%

Table A-11 (Continued)

		Montana	Nebraska	Nevada	NH	New Jersey	New Mexico	New York	North Carolina	North Dakota	Ohio	Oklahoma	Oregon	Pennsylvania
Cumulative Installations	2007	1	1	15	1	69	9	32	3	1	2	1	2	9
	2008	1	1	79	1	103	12	128	6	1	6	1	6	10
	2009	1	1	107	1	140	12	134	10	1	6	1	9	23
	2010	1	1	109	4	194	12	140	22	1	6	1	76	37
	2011	1	1	140	8	253	12	146	22	1	6	1	112	58
	2012	1	1	143	16	321	14	153	76	1	6	1	150	96
	2013	1	1	175	22	405	16	159	76	1	6	1	193	166
	2014	1	1	179	33	502	21	160	76	1	9	1	258	290
	2015	1	1	203	33	614	25	161	154	1	14	1	333	343
Annual Installations	2007	0	0	7	1	29	3	10	1	0	1	0	1	3
	2008	0	0	64	0	34	3	95	3	0	4	0	4	0
	2009	0	0	28	0	37	0	6	4	0	0	0	3	13
	2010	0	0	2	3	55	0	6	12	0	0	0	67	14
	2011	0	0	30	4	59	0	6	0	0	0	0	36	21
	2012	0	0	3	8	68	2	6	54	0	0	0	38	38
	2013	0	0	32	6	83	2	6	0	0	0	0	43	70
	2014	0	0	4	11	98	5	1	0	0	3	1	65	124
	2015	0	0	24	0	111	4	1	77	0	5	0	75	53
Installers Required	2007	0	0	56	4	213	24	71	7	0	4	0	4	19
	2008	3	0	12	0	243	23	43	22	0	26	0	30	3
	2009	0	0	198	2	255	1	42	27	0	0	0	18	91
	2010	0	0	16	19	366	0	41	82	0	0	0	451	93
	2011	0	0	189	27	366	0	39	0	0	0	0	226	133
	2012	0	0	17	44	397	9	37	310	0	0	0	220	221
	2013	0	0	172	30	444	11	34	2	0	1	0	231	373
	2014	0	0	18	54	475	25	6	2	0	16	3	315	606
	2015	0	0	107	2	492	20	5	341	0	24	1	332	233
Market Penetration	2007	0%	0%	0%	0%	1%	0%	0%	0%	0%	0%	0%	0%	0%
	2008	0%	0%	1%	0%	1%	0%	1%	0%	0%	0%	0%	0%	0%
	2009	0%	0%	2%	0%	2%	0%	1%	0%	0%	0%	0%	0%	0%
	2010	0%	0%	2%	0%	2%	0%	1%	0%	0%	0%	0%	1%	0%
	2011	0%	0%	2%	0%	3%	0%	1%	0%	0%	0%	0%	2%	0%
	2012	0%	0%	2%	1%	3%	0%	1%	0%	0%	0%	0%	2%	1%
	2013	0%	0%	2%	1%	4%	0%	1%	0%	0%	0%	0%	3%	1%
	2014	0%	0%	2%	2%	5%	0%	1%	0%	0%	0%	0%	3%	2%
	2015	0%	0%	2%	2%	5%	1%	1%	1%	0%	0%	0%	4%	2%

		Rhode Island	South Carolina	South Dakota	Tennessee	Texas	Utah	Vermont	Virginia	Washington	DC	West Virginia	Wisconsin	Wyoming
Cumulative Installations	2007	1	2	1	1	6	1	2	1	6	1	1	6	1
	2008	1	2	1	6	8	1	2	1	8	1	1	10	1
	2009	1	2	1	6	12	1	2	1	9	1	1	10	1
	2010	1	2	1	6	20	1	2	1	11	3	1	10	1
	2011	1	2	1	6	23	1	2	1	11	3	1	10	1
	2012	1	2	1	6	27	1	3	1	13	7	1	10	1
	2013	1	2	1	6	32	1	3	1	15	7	1	10	1
	2014	2	2	1	6	45	1	3	2	21	7	1	15	1
	2015	2	3	1	6	63	1	4	3	21	7	1	17	1
Annual Installations	2007	1	1	0	0	1	1	1	1	3	1	0	2	0
	2008	0	0	0	5	2	0	0	1	2	0	0	4	0
	2009	0	0	0	0	4	0	0	0	1	0	0	0	0
	2010	0	0	0	0	8	0	0	0	2	1	0	0	0
	2011	0	0	0	0	3	0	0	0	0	0	0	0	0
	2012	0	0	0	0	4	0	0	0	2	4	0	0	0
	2013	0	0	0	0	4	0	0	0	2	0	0	0	0
	2014	1	0	0	0	13	0	1	0	6	0	0	5	0
	2015	0	1	0	0	18	0	0	2	0	0	0	2	0
Installers Required	2007	4	7	0	0	8	4	7	4	22	4	0	15	0
	2008	0	0	0	37	16	0	1	4	17	1	0	29	0
	2009	0	0	0	0	26	0	0	0	5	2	0	0	0
	2010	1	0	0	0	51	0	0	0	12	8	0	0	0
	2011	1	0	0	0	20	0	1	0	0	0	0	0	0
	2012	1	0	0	0	25	0	1	0	12	24	0	0	0
	2013	1	0	0	0	24	0	1	0	13	0	0	2	0
	2014	5	1	0	0	65	0	3	1	30	1	0	24	1
	2015	0	3	0	1	81	0	1	8	0	0	0	7	1
Market Penetration	2007	0%	0%	0%	0%	0%	0%	0%	0%	0%	0%	0%	0%	0%
	2008	0%	0%	0%	0%	0%	0%	0%	0%	0%	0%	0%	0%	0%
	2009	0%	0%	0%	0%	0%	0%	0%	0%	0%	0%	0%	0%	0%
	2010	0%	0%	0%	0%	0%	0%	0%	0%	0%	0%	0%	0%	0%
	2011	0%	0%	0%	0%	0%	0%	0%	0%	0%	0%	0%	0%	0%
	2012	0%	0%	0%	0%	0%	0%	0%	0%	0%	0%	0%	0%	0%
	2013	0%	0%	0%	0%	0%	0%	0%	0%	0%	0%	0%	0%	0%
	2014	0%	0%	0%	0%	0%	0%	0%	0%	0%	0%	0%	0%	0%
	2015	0%	0%	0%	0%	0%	0%	0%	0%	0%	0%	0%	0%	0%

Table A-12. State-by-State Results for the Base Case, with SAI System Pricing

		Alabama	Alaska	Arizona	Arkansas	California	Colorado	Connecticut	Delaware	Florida	Georgia	Hawaii	Idaho	Illinois
Cumulative Installations	2007	1	1	14	1	499	20	3	1	1	1	6	1	1
	2008	1	1	41	1	1,220	34	3	11	2	1	10	1	1
	2009	1	1	72	1	1,220	35	8	11	3	1	10	1	1
	2010	1	1	122	1	1,220	36	11	11	4	3	10	1	1
	2011	1	1	187	1	1,256	73	13	11	15	6	10	1	1
	2012	1	1	268	1	1,524	75	19	11	24	9	11	1	1
	2013	1	1	313	1	1,886	77	26	11	32	14	12	1	1
	2014	1	1	360	1	2,890	78	82	16	48	24	22	1	6
	2015	3	1	408	1	3,202	146	108	18	69	31	27	2	14
Annual Installations	2007	0	0	5	0	166	6	1	1	1	0	3	0	0
	2008	0	0	27	0	721	14	0	10	1	0	4	0	0
	2009	0	0	31	0	0	1	5	0	1	1	0	0	0
	2010	0	0	50	0	0	1	4	0	1	2	0	0	0
	2011	0	0	65	0	36	37	2	0	11	3	0	0	0
	2012	0	0	81	0	269	2	6	0	9	3	0	0	0
	2013	0	0	45	0	361	2	7	0	8	4	2	0	0
	2014	0	0	47	0	1,005	2	56	5	16	11	10	0	5
	2015	2	0	49	0	311	67	26	1	21	6	4	1	8
Installers Required	2007	0	0	35	0	1,232	48	7	4	4	0	23	0	0
	2008	0	0	171	0	5,192	102	2	75	6	0	32	2	0
	2009	0	0	214	0	0	5	33	0	6	4	0	1	0
	2010	0	0	339	0	0	5	25	0	6	16	0	0	0
	2011	0	0	407	0	225	235	10	0	69	16	0	0	0
	2012	0	0	467	0	1,557	9	34	24	52	20	1	0	0
	2013	0	0	239	0	1,927	9	38	0	44	23	9	0	2
	2014	0	0	228	0	4,898	8	274	49	80	51	48	2	25
	2015	9	0	214	0	1,374	298	114	6	91	28	20	3	37
Market Penetration	2007	0%	0%	0%	0%	1%	0%	0%	0%	0%	0%	0%	0%	0%
	2008	0%	0%	0%	0%	2%	0%	0%	1%	0%	0%	1%	0%	0%
	2009	0%	0%	1%	0%	2%	0%	0%	1%	0%	0%	0%	0%	0%
	2010	0%	0%	1%	0%	2%	0%	0%	1%	0%	0%	0%	0%	0%
	2011	0%	0%	1%	0%	2%	1%	0%	1%	0%	0%	0%	0%	0%
	2012	0%	0%	2%	0%	2%	1%	0%	1%	0%	0%	0%	0%	0%
	2013	0%	0%	2%	0%	3%	1%	0%	1%	0%	0%	0%	0%	0%
	2014	0%	0%	2%	0%	4%	1%	1%	1%	0%	0%	1%	0%	0%
	2015	0%	0%	2%	0%	4%	1%	2%	1%	0%	0%	1%	0%	0%

		Indiana	Iowa	Kansas	Kentucky	Louisiana	Maine	Maryland	Mass.	Michigan	Minnesota	Mississippi	Missouri
Cumulative Installations	2007	1	1	1	1	1	2	2	7	1	1	1	1
	2008	1	1	1	2	1	3	3	9	1	1	1	1
	2009	1	1	1	2	1	41	15	10	1	1	1	1
	2010	1	1	1	2	1	119	15	12	1	1	1	1
	2011	1	1	1	2	1	119	35	31	1	1	1	1
	2012	1	1	1	2	1	119	35	64	2	1	2	1
	2013	1	1	1	2	1	129	89	94	5	2	2	1
	2014	1	1	1	3	3	176	89	213	12	7	3	1
	2015	1	1	1	5	4	216	212	283	16	9	5	1
Annual Installations	2007	0	0	0	0	0	1	1	2	1	0	0	0
	2008	0	0	0	1	0	1	1	2	0	0	0	0
	2009	0	0	0	0	0	38	13	2	0	0	0	0
	2010	0	0	0	0	0	78	0	2	0	0	0	0
	2011	0	0	0	0	0	0	20	19	0	0	0	0
	2012	0	0	0	0	0	0	0	33	1	0	1	0
	2013	0	0	0	0	0	9	53	30	3	1	0	0
	2014	0	0	0	1	2	47	0	119	7	5	1	0
	2015	0	0	0	2	1	40	123	71	4	2	2	1
Installers Required	2007	0	0	0	0	0	7	7	16	4	0	0	0
	2008	0	0	0	10	0	4	4	13	0	2	0	0
	2009	0	0	0	0	0	268	89	13	0	0	1	0
	2010	0	0	0	0	0	526	0	12	0	0	2	0
	2011	0	0	0	0	0	0	125	118	2	0	2	0
	2012	0	0	0	0	2	0	0	191	5	2	3	0
	2013	0	0	0	0	3	50	284	159	13	6	3	0
	2014	0	1	1	4	10	230	0	580	36	22	5	0
	2015	0	2	1	10	4	176	545	312	18	10	8	4
Market Penetration	2007	0%	0%	0%	0%	0%	0%	0%	0%	0%	0%	0%	0%
	2008	0%	0%	0%	0%	0%	0%	0%	0%	0%	0%	0%	0%
	2009	0%	0%	0%	0%	0%	2%	0%	0%	0%	0%	0%	0%
	2010	0%	0%	0%	0%	0%	7%	0%	0%	0%	0%	0%	0%
	2011	0%	0%	0%	0%	0%	7%	0%	0%	0%	0%	0%	0%
	2012	0%	0%	0%	0%	0%	6%	0%	1%	0%	0%	0%	0%
	2013	0%	0%	0%	0%	0%	6%	1%	1%	0%	0%	0%	0%
	2014	0%	0%	0%	0%	0%	8%	1%	2%	0%	0%	0%	0%
	2015	0%	0%	0%	0%	0%	10%	2%	3%	0%	0%	0%	0%

Table A-12 (Continued)

		Montana	Nebraska	Nevada	NH	New Jersey	New Mexico	New York	North Carolina	North Dakota	Ohio	Oklahoma	Oregon	Pennsylvania
Cumulative Installations	2007	1	1	15	1	69	9	32	3	1	2	1	2	9
	2008	1	1	79	1	103	13	128	6	1	7	1	6	10
	2009	1	1	107	2	140	14	134	13	1	7	1	9	23
	2010	1	1	109	6	194	14	140	55	1	7	1	118	37
	2011	1	1	140	8	253	15	146	55	1	7	1	167	58
	2012	1	1	143	16	321	22	153	76	1	9	1	232	96
	2013	1	1	175	22	405	32	159	76	1	12	2	295	166
	2014	2	1	179	33	568	73	160	76	1	23	3	400	290
	2015	2	1	203	34	736	110	161	154	1	40	4	508	343
Annual Installations	2007	0	0	7	1	29	3	10	1	0	1	0	1	3
	2008	0	0	64	0	34	4	95	3	0	5	0	4	1
	2009	0	0	28	1	37	1	6	6	0	0	0	3	13
	2010	0	0	2	4	55	0	6	43	0	0	0	109	14
	2011	0	0	30	3	59	1	6	0	0	0	0	50	21
	2012	0	0	3	8	68	7	6	20	0	2	0	64	38
	2013	0	0	32	6	83	10	6	0	0	3	0	63	70
	2014	1	0	4	11	164	40	1	0	0	10	2	106	124
	2015	1	0	24	1	167	38	1	77	0	17	1	108	53
Installers Required	2007	0	0	56	4	213	24	71	7	0	4	0	4	19
	2008	3	0	12	0	243	32	43	22	0	34	0	30	6
	2009	0	0	198	5	255	4	42	45	0	0	0	18	88
	2010	0	0	16	28	366	0	41	287	0	0	1	730	93
	2011	0	0	189	16	366	8	39	0	0	1	1	312	133
	2012	0	0	17	44	397	41	37	118	0	11	3	373	221
	2013	0	0	172	30	444	54	34	2	0	19	2	334	373
	2014	3	1	18	54	798	197	6	2	0	51	9	514	606
	2015	3	1	107	6	738	166	5	341	0	76	4	477	233
Market Penetration	2007	0%	0%	0%	0%	1%	0%	0%	0%	0%	0%	0%	0%	0%
	2008	0%	0%	1%	0%	1%	0%	1%	0%	0%	0%	0%	0%	0%
	2009	0%	0%	2%	0%	2%	0%	1%	0%	0%	0%	0%	0%	0%
	2010	0%	0%	2%	0%	2%	0%	1%	0%	0%	0%	0%	2%	0%
	2011	0%	0%	2%	0%	3%	0%	1%	0%	0%	0%	0%	2%	0%
	2012	0%	0%	2%	1%	3%	1%	1%	0%	0%	0%	0%	3%	1%
	2013	0%	0%	2%	1%	4%	1%	1%	0%	0%	0%	0%	4%	1%
	2014	0%	0%	2%	2%	5%	2%	1%	0%	0%	0%	0%	5%	2%
	2015	0%	0%	2%	2%	6%	2%	1%	1%	0%	0%	0%	6%	2%

		Rhode Island	South Carolina	South Dakota	Tennessee	Texas	Utah	Vermont	Virginia	Washington	DC	West Virginia	Wisconsin	Wyoming
Cumulative Installations	2007	1	2	1	1	6	1	2	1	6	1	1	6	1
	2008	1	2	1	7	9	1	2	1	9	1	1	11	1
	2009	1	2	1	7	17	1	2	1	11	1	1	11	1
	2010	1	2	1	7	41	1	.3	1	17	4	1	13	1
	2011	1	2	1	7	41	1	3	1	17	4	1	13	1
	2012	2	3	1	7	52	1	3	2	24	7	1	16	1
	2013	2	4	1	7	64	1	4	4	39	8	1	23	1
	2014	4	8	1	10	112	1	6	10	79	9	2	37	1
	2015	4	10	1	16	219	2	8	14	79	9	3	46	1
Annual Installations	2007	1	1	0	0	1	1	1	1	3	1	0	2	0
	2008	0	0	0	6	3	0	0	1	3	0	0	5	0
	2009	0	0	0	0	8	0	0	0	2	0	0	0	0
	2010	1	0	0	0	25	0	1	0	6	3	0	2	0
	2011	0	0	0	0	0	0	0	0	0	0	0	0	0
	2012	0	1	0	0	10	0	1	1	7	2	0	3	0
	2013	1	1	0	0	12	0	1	2	16	2	0	6	0
	2014	1	3	0	4	48	0	2	5	40	0	1	15	0
	2015	1	2	0	5	107	2	2	4	0	0	1	9	0
Installers Required	2007	4	7	0	0	8	4	7	4	22	4	0	15	0
	2008	0	0	0	43	24	0	2	4	23	1	0	37	0
	2009	1	0	0	0	53	0	0	0	16	2	0	0	0
	2010	5	0	0	0	165	0	4	0	37	20	0	13	0
	2011	0	2	0	0	0	0	0	0	1	0	0	0	1
	2012	2	5	0	0	60	0	4	6	39	13	0	19	1
	2013	3	6	0	0	64	0	5	9	83	13	0	33	1
	2014	5	17	0	18	234	0	10	27	195	0	5	71	1
	2015	4	8	0	24	474	7	7	19	0	0	5	41	1
Market Penetration	2007	0%	0%	0%	0%	0%	0%	0%	0%	0%	0%	0%	0%	0%
	2008	0%	0%	0%	0%	0%	0%	0%	0%	0%	0%	0%	0%	0%
	2009	0%	0%	0%	0%	0%	0%	0%	0%	0%	0%	0%	0%	0%
	2010	0%	0%	0%	0%	0%	0%	0%	0%	0%	0%	0%	0%	0%
	2011	0%	0%	0%	0%	0%	0%	0%	0%	0%	0%	0%	0%	0%
	2012	0%	0%	0%	0%	0%	0%	0%	0%	0%	0%	0%	0%	0%
	2013	0%	0%	0%	0%	0%	0%	0%	0%	0%	1%	0%	0%	0%
	2014	0%	0%	0%	0%	0%	0%	1%	0%	1%	1%	0%	0%	0%
	2015	0%	0%	0%	0%	0%	0%	1%	0%	1%	0%	0%	0%	0%

Table A-13. State-by-State Results for the Focused Policy Case, BAU System Pricing

		Alabama	Alaska	Arizona	Arkansas	California	Colorado	Connecticut	Delaware	Florida	Georgia	Hawaii	Idaho	Illinois
Cumulative Installations	2007	1	1	14	1	499	20	3	1	1	1	6	1	1
	2008	1	1	41	1	1,051	34	3	10	2	1	10	1	1
	2009	1	1	72	1	1,105	35	6	10	3	1	10	1	1
	2010	1	1	122	1	1,105	36	8	10	4	1	10	1	1
	2011	1	1	187	1	1,603	73	10	10	5	1	10	1	1
	2012	1	1	268	1	2,364	75	15	10	10	1	10	1	1
	2013	1	1	313	1	2,713	77	20	10	13	1	12	1	1
	2014	5	1	360	1	4,317	78	30	16	22	9	15	1	1
	2015	11	1	408	1	5,314	120	35	18	44	11	18	1	2
Annual Installations	2007	0	0	5	0	166	6	1	1	1	0	3	0	0
	2008	0	0	27	0	552	14	0	9	1	0	4	0	0
	2009	0	0	31	0	55	1	3	0	1	0	0	0	0
	2010	0	0	50	0	0	1	1	0	1	0	0	0	0
	2011	0	0	65	0	497	37	3	0	1	0	0	0	0
	2012	0	0	81	0	761	2	5	0	5	0	0	0	0
	2013	0	0	45	0	349	2	5	0	3	1	2	0	0
	2014	4	0	47	0	1,604	2	10	6	9	7	3	0	0
	2015	6	0	49	0	997	42	5	1	22	3	3	0	1
Installers Required	2007	0	0	35	0	1,232	48	7	4	4	0	23	0	0
	2008	0	0	171	0	3,973	102	2	64	6	0	27	1	0
	2009	0	0	214	0	379	5	24	0	6	0	0	1	0
	2010	0	0	339	0	0	5	9	0	6	0	0	0	0
	2011	0	0	407	0	3,112	235	17	0	8	0	0	0	0
	2012	0	0	467	0	4,413	9	28	0	28	0	0	0	0
	2013	2	0	239	0	1,861	9	27	0	17	5	11	0	0
	2014	18	1	228	0	7,821	8	48	49	43	35	17	0	0
	2015	28	1	214	0	4,401	184	22	6	98	12	13	0	5
Market Penetration	2007	0%	0%	0%	0%	1%	0%	0%	0%	0%	0%	0%	0%	0%
	2008	0%	0%	0%	0%	2%	0%	0%	1%	0%	0%	0%	0%	0%
	2009	0%	0%	1%	0%	2%	0%	0%	1%	0%	0%	0%	0%	0%
	2010	0%	0%	1%	0%	2%	0%	0%	1%	0%	0%	0%	0%	0%
	2011	0%	0%	1%	0%	2%	1%	0%	1%	0%	0%	0%	0%	0%
	2012	0%	0%	2%	0%	3%	1%	0%	1%	0%	0%	0%	0%	0%
	2013	0%	0%	2%	0%	4%	1%	0%	1%	0%	0%	0%	0%	0%
	2014	0%	0%	2%	0%	6%	1%	1%	1%	0%	0%	1%	0%	0%
	2015	0%	0%	2%	0%	7%	1%	1%	1%	0%	0%	1%	0%	0%

		Indiana	Iowa	Kansas	Kentucky	Louisiana	Maine	Maryland	Mass.	Michigan	Minnesota	Mississippi	Missouri
Cumulative Installations	2007	1	1	1	1	1	2	2	7	1	1	1	1
	2008	1	1	1	2	1	2	3	9	1	1	1	1
	2009	1	1	1	2	1	25	15	10	1	1	1	1
	2010	1	1	1	2	1	79	24	12	1	1	1	1
	2011	1	1	1	2	1	79	24	25	1	1	1	1
	2012	1	1	1	2	1	84	59	36	1	1	1	1
	2013	1	1	1	2	1	108	59	48	1	1	1	1
	2014	1	1	1	2	1	139	149	70	2	1	2	1
	2015	1	1	1	2	1	168	149	81	2	1	2	1
Annual Installations	2007	0	0	0	0	0	1	1	2	1	0	0	0
	2008	0	0	0	1	0	0	1	2	0	0	0	0
	2009	0	0	0	0	0	22	13	2	0	0	0	0
	2010	0	0	0	0	0	55	9	2	0	0	0	0
	2011	0	0	0	0	0	0	0	13	0	0	0	0
	2012	0	0	0	0	0	5	35	11	0	0	0	0
	2013	0	0	0	0	0	24	0	12	0	0	0	0
	2014	0	0	0	0	0	30	90	22	1	0	1	0
	2015	0	0	0	0	0	29	0	11	1	0	1	0
Installers Required	2007	0	0	0	0	0	7	7	16	4	0	0	0
	2008	0	0	0	8	0	3	4	13	0	2	0	0
	2009	0	0	0	0	0	155	88	13	0	0	0	0
	2010	0	0	0	0	0	367	57	12	0	0	0	0
	2011	0	0	0	0	0	0	0	81	0	0	0	0
	2012	0	0	0	0	0	29	202	63	0	0	2	0
	2013	0	0	0	0	0	129	0	62	0	0	1	0
	2014	0	0	0	0	0	148	441	109	5	0	3	0
	2015	0	0	0	0	0	128	0	49	3	0	3	0
Market Penetration	2007	0%	0%	0%	0%	0%	0%	0%	0%	0%	0%	0%	0%
	2008	0%	0%	0%	0%	0%	0%	0%	0%	0%	0%	0%	0%
	2009	0%	0%	0%	0%	0%	1%	0%	0%	0%	0%	0%	0%
	2010	0%	0%	0%	0%	0%	5%	0%	0%	0%	0%	0%	0%
	2011	0%	0%	0%	0%	0%	4%	0%	0%	0%	0%	0%	0%
	2012	0%	0%	0%	0%	0%	4%	1%	0%	0%	0%	0%	0%
	2013	0%	0%	0%	0%	0%	5%	1%	1%	0%	0%	0%	0%
	2014	0%	0%	0%	0%	0%	7%	1%	1%	0%	0%	0%	0%
	2015	0%	0%	0%	0%	0%	8%	1%	1%	0%	0%	0%	0%

Table A-13 (Continued)

		Montana	Nebraska	Nevada	NH	New Jersey	New Mexico	New York	North Carolina	North Dakota	Ohio	Oklahoma	Oregon	Pennsylvania
Cumulative Installations	2007	1	1	15	1	89	9	32	3	1	2	1	2	9
	2008	1	1	79	1	103	12	128	6	1	6	1	6	10
	2009	1	1	107	1	140	12	134	10	1	6	1	9	23
	2010	1	1	109	4	194	12	140	22	1	6	1	76	37
	2011	1	1	140	8	253	13	146	22	1	6	1	112	58
	2012	1	1	143	16	321	17	153	76	1	6	1	152	96
	2013	1	1	175	22	405	21	159	76	1	6	1	199	166
	2014	1	1	179	33	502	29	239	76	1	9	1	271	290
	2015	1	1	203	33	614	35	280	154	1	14	1	352	343
Annual Installations	2007	0	0	7	1	29	3	10	1	0	1	0	1	3
	2008	0	0	64	0	34	3	95	3	0	4	0	4	0
	2009	0	0	28	0	37	0	6	4	0	0	0	3	13
	2010	0	0	2	3	55	0	6	12	0	0	0	67	14
	2011	0	0	30	4	59	1	6	0	0	0	0	36	21
	2012	0	0	3	8	68	4	6	54	0	0	0	40	38
	2013	0	0	32	6	83	4	6	0	0	0	0	47	70
	2014	0	0	4	11	98	8	80	0	0	3	1	72	124
	2015	0	0	24	0	111	6	41	77	0	5	0	81	53
Installers Required	2007	0	0	56	4	213	24	71	7	0	4	0	4	19
	2008	3	0	12	0	243	23	43	22	0	26	0	30	3
	2009	0	0	198	2	255	1	42	27	0	0	0	18	91
	2010	0	0	16	19	366	0	41	82	0	0	0	451	93
	2011	0	0	189	27	366	5	39	0	0	0	0	226	133
	2012	0	0	17	44	397	25	37	310	0	0	0	230	221
	2013	0	0	172	30	444	21	34	2	0	1	0	249	373
	2014	0	0	18	54	475	38	390	2	0	16	3	352	606
	2015	0	0	107	2	492	26	182	341	0	24	1	357	233
Market Penetration	2007	0%	0%	0%	0%	1%	0%	0%	0%	0%	0%	0%	0%	0%
	2008	0%	0%	1%	0%	1%	0%	1%	0%	0%	0%	0%	0%	0%
	2009	0%	0%	2%	0%	2%	0%	1%	0%	0%	0%	0%	0%	0%
	2010	0%	0%	2%	0%	2%	0%	1%	0%	0%	0%	0%	1%	0%
	2011	0%	0%	2%	0%	3%	0%	1%	0%	0%	0%	0%	2%	0%
	2012	0%	0%	2%	1%	3%	0%	1%	0%	0%	0%	0%	2%	1%
	2013	0%	0%	2%	1%	4%	1%	1%	0%	0%	0%	0%	3%	1%
	2014	0%	0%	2%	2%	5%	1%	1%	0%	0%	0%	0%	3%	2%
	2015	0%	0%	2%	2%	5%	1%	1%	1%	0%	0%	0%	4%	2%

		Rhode Island	South Carolina	South Dakota	Tennessee	Texas	Utah	Vermont	Virginia	Washington	DC	West Virginia	Wisconsin	Wyoming
Cumulative Installations	2007	1	2	1	1	6	1	2	1	6	1	1	6	1
	2008	1	2	1	7	8	1	2	1	8	1	1	10	1
	2009	1	2	1	7	12	1	2	1	9	1	1	10	1
	2010	1	2	1	7	20	1	2	1	11	3	1	10	1
	2011	1	2	1	7	23	1	2	1	14	3	1	10	1
	2012	2	2	1	7	27	1	3	1	19	7	1	14	1
	2013	3	2	1	7	32	1	4	1	24	10	1	18	1
	2014	4	7	1	7	56	1	5	2	32	10	1	28	1
	2015	5	8	1	7	117	1	6	3	32	10	1	31	1
Annual Installations	2007	1	1	0	0	1	1	1	1	3	1	0	2	0
	2008	0	0	0	7	2	0	0	1	2	0	0	4	0
	2009	0	0	0	0	4	0	0	0	1	0	0	0	0
	2010	0	0	0	0	8	0	0	0	2	1	0	0	0
	2011	0	0	0	0	3	0	0	0	3	0	0	0	0
	2012	1	0	0	0	4	0	1	0	5	4	0	4	0
	2013	1	0	0	0	4	0	1	0	5	3	0	4	0
	2014	1	5	0	0	25	0	1	0	8	0	0	10	0
	2015	1	1	0	0	60	0	1	2	0	0	0	3	0
Installers Required	2007	4	7	0	0	8	4	7	4	22	4	0	15	0
	2008	0	0	0	50	16	0	1	4	17	1	0	29	0
	2009	0	0	0	0	26	0	0	0	5	2	0	0	0
	2010	1	0	0	0	51	0	0	0	12	8	0	0	0
	2011	2	0	0	0	20	0	2	0	18	0	0	0	0
	2012	5	0	0	0	25	0	4	0	28	24	0	21	0
	2013	3	0	0	0	24	0	3	0	27	16	0	23	0
	2014	6	24	0	0	121	0	6	1	40	0	0	48	1
	2015	3	4	0	0	266	0	2	8	0	0	0	14	1
Market Penetration	2007	0%	0%	0%	0%	0%	0%	0%	0%	0%	0%	0%	0%	0%
	2008	0%	0%	0%	0%	0%	0%	0%	0%	0%	0%	0%	0%	0%
	2009	0%	0%	0%	0%	0%	0%	0%	0%	0%	0%	0%	0%	0%
	2010	0%	0%	0%	0%	0%	0%	0%	0%	0%	0%	0%	0%	0%
	2011	0%	0%	0%	0%	0%	0%	0%	0%	0%	0%	0%	0%	0%
	2012	0%	0%	0%	0%	0%	0%	0%	0%	0%	0%	0%	0%	0%
	2013	0%	0%	0%	0%	0%	0%	0%	0%	0%	1%	0%	0%	0%
	2014	0%	0%	0%	0%	0%	0%	0%	0%	0%	1%	0%	0%	0%
	2015	0%	0%	0%	0%	0%	0%	1%	0%	0%	1%	0%	0%	0%

Table A-14. State-by-State Results for the Focused Policy Case, SAI System Pricing

		Alabama	Alaska	Arizona	Arkansas	California	Colorado	Connecticut	Delaware	Florida	Georgia	Hawaii	Idaho	Illinois
Cumulative Installations	2007	1	1	14	1	499	20	3	1	1	1	6	1	1
	2008	1	1	41	1	1,220	34	3	11	2	1	10	1	1
	2009	1	1	72	1	1,441	35	8	11	3	1	10	1	1
	2010	1	1	122	1	1,771	36	11	11	4	3	10	1	1
	2011	2	1	187	1	2,757	73	18	11	15	6	12	1	1
	2012	4	1	268	1	4,009	75	27	11	31	9	16	1	1
	2013	10	1	313	1	5,210	77	50	11	63	14	21	1	1
	2014	22	2	360	1	7,961	95	176	16	135	24	39	1	6
	2015	35	2	408	1	10,772	179	246	28	198	31	70	2	16
Annual Installations	2007	0	0	5	0	166	6	1	1	1	0	3	0	0
	2008	0	0	27	0	721	14	0	10	1	0	4	0	0
	2009	0	0	31	0	222	1	5	0	1	1	0	0	0
	2010	0	0	50	0	330	1	4	0	1	2	0	0	0
	2011	1	0	65	0	985	37	6	0	11	3	2	0	0
	2012	3	0	81	0	1,252	2	9	0	16	3	4	0	0
	2013	5	0	45	0	1,201	2	24	0	32	4	5	0	0
	2014	12	0	47	0	2,751	18	126	5	72	11	18	0	5
	2015	13	1	49	0	2,810	84	70	12	64	6	31	1	10
Installers Required	2007	0	0	35	0	1,232	48	7	4	4	0	23	0	0
	2008	0	0	171	0	5,192	102	2	75	6	0	32	2	0
	2009	0	0	214	0	1,541	5	33	0	6	4	0	1	0
	2010	2	0	339	0	2,215	5	25	0	6	16	0	0	0
	2011	6	2	407	0	6,163	235	39	2	70	16	12	0	0
	2012	15	1	467	0	7,258	9	52	22	93	20	20	0	0
	2013	28	2	239	0	6,407	9	127	0	169	23	26	0	2
	2014	60	2	228	0	13,413	87	612	49	352	51	86	2	25
	2015	58	3	214	1	12,406	372	311	53	281	29	138	3	46
Market Penetration	2007	0%	0%	0%	0%	1%	0%	0%	0%	0%	0%	0%	0%	0%
	2008	0%	0%	0%	0%	2%	0%	0%	1%	0%	0%	1%	0%	0%
	2009	0%	0%	1%	0%	2%	0%	0%	1%	0%	0%	0%	0%	0%
	2010	0%	0%	1%	0%	3%	0%	0%	1%	0%	0%	0%	0%	0%
	2011	0%	0%	1%	0%	4%	1%	0%	1%	0%	0%	1%	0%	0%
	2012	0%	0%	2%	0%	6%	1%	1%	1%	0%	0%	1%	0%	0%
	2013	0%	0%	2%	0%	7%	1%	1%	1%	0%	0%	1%	0%	0%
	2014	0%	0%	2%	0%	10%	1%	3%	1%	0%	0%	1%	0%	0%
	2015	0%	0%	2%	0%	13%	1%	4%	1%	0%	0%	2%	0%	0%

		Indiana	Iowa	Kansas	Kentucky	Louisiana	Maine	Maryland	Mass.	Michigan	Minnesota	Mississippi	Missouri
Cumulative Installations	2007	1	1	1	1	1	2	2	7	1	1	1	1
	2008	1	1	1	2	1	3	3	9	1	1	1	1
	2009	1	1	1	2	1	41	15	10	1	1	1	1
	2010	1	1	1	2	1	119	15	12	1	1	1	1
	2011	1	1	1	2	1	119	35	41	1	1	1	1
	2012	1	1	1	2	1	126	35	80	2	1	2	1
	2013	1	1	1	2	1	165	89	136	5	2	2	1
	2014	1	1	1	3	3	213	89	405	22	12	8	1
	2015	11	1	1	5	4	266	212	573	32	16	14	1
Annual Installations	2007	0	0	0	0	0	1	1	2	1	0	0	0
	2008	0	0	0	1	0	1	1	2	0	0	0	0
	2009	0	0	0	0	0	38	13	2	0	0	0	0
	2010	0	0	0	0	0	78	0	2	0	0	0	0
	2011	0	0	0	0	0	0	20	29	0	0	0	0
	2012	0	0	0	0	0	7	0	39	1	0	1	0
	2013	0	0	0	0	0	39	53	56	3	1	1	0
	2014	0	0	0	1	2	47	0	268	17	10	6	0
	2015	11	0	0	2	1	53	123	168	10	4	6	1
Installers Required	2007	0	0	0	0	0	7	7	16	4	0	0	0
	2008	0	0	0	10	0	4	4	13	0	2	0	0
	2009	0	0	0	0	0	268	89	13	0	0	1	0
	2010	0	0	0	0	0	526	0	12	0	0	2	0
	2011	0	0	0	0	0	0	125	179	2	0	2	0
	2012	0	0	0	0	2	42	0	226	5	2	3	0
	2013	0	0	0	0	3	208	284	301	16	6	3	0
	2014	0	1	1	4	10	231	0	1,308	83	47	29	0
	2015	48	2	1	10	4	233	545	743	43	19	26	4
Market Penetration	2007	0%	0%	0%	0%	0%	0%	0%	0%	0%	0%	0%	0%
	2008	0%	0%	0%	0%	0%	0%	0%	0%	0%	0%	0%	0%
	2009	0%	0%	0%	0%	0%	2%	0%	0%	0%	0%	0%	0%
	2010	0%	0%	0%	0%	0%	7%	0%	0%	0%	0%	0%	0%
	2011	0%	0%	0%	0%	0%	7%	0%	0%	0%	0%	0%	0%
	2012	0%	0%	0%	0%	0%	7%	0%	1%	0%	0%	0%	0%
	2013	0%	0%	0%	0%	0%	8%	1%	1%	0%	0%	0%	0%
	2014	0%	0%	0%	0%	0%	10%	1%	4%	0%	0%	0%	0%
	2015	0%	0%	0%	0%	0%	12%	2%	6%	0%	0%	0%	0%

Table A-14 (Continued)

		Montana	Nebraska	Nevada	NH	New Jersey	New Mexico	New York	North Carolina	North Dakota	Ohio	Oklahoma	Oregon	Pennsylvania
Cumulative Installations	2007	1	1	15	1	69	9	32	3	1	2	1	2	9
	2008	1	1	79	1	103	13	128	6	1	7	1	6	10
	2009	1	1	107	2	140	14	134	13	1	7	1	9	23
	2010	1	1	109	6	194	14	140	55	1	7	1	118	37
	2011	1	1	140	8	262	18	146	55	1	7	1	173	58
	2012	1	1	143	22	386	28	210	76	1	9	1	244	101
	2013	1	1	175	22	530	40	373	76	1	12	2	319	166
	2014	3	1	179	33	530	112	698	76	1	23	3	504	290
	2015	5	1	276	50	614	173	838	154	1	51	4	830	511
Annual Installations	2007	0	0	7	1	29	3	10	1	0	1	0	1	3
	2008	0	0	64	0	34	4	95	3	0	5	0	4	1
	2009	0	0	28	1	37	1	6	6	0	0	0	3	13
	2010	0	0	2	4	55	0	6	43	0	0	0	109	14
	2011	0	0	30	3	68	5	6	0	0	0	0	54	21
	2012	0	0	3	13	124	10	64	20	0	2	0	72	43
	2013	0	0	32	0	145	12	164	0	0	3	0	75	65
	2014	2	1	4	11	0	72	325	0	0	10	2	185	124
	2015	2	0	98	18	83	62	140	77	0	28	1	326	221
Installers Required	2007	0	0	56	4	213	24	71	7	0	4	0	4	19
	2008	3	0	12	0	243	32	43	22	0	34	0	30	6
	2009	0	0	198	5	255	4	42	45	0	0	0	18	88
	2010	0	0	16	28	366	0	41	287	0	0	1	734	93
	2011	0	0	189	16	424	30	39	0	0	1	1	340	133
	2012	0	0	17	77	717	55	368	118	0	11	3	414	252
	2013	0	0	172	0	771	63	873	2	0	19	2	402	344
	2014	10	3	18	53	0	350	1,584	2	0	51	9	902	606
	2015	7	2	431	77	367	273	618	341	0	123	4	1,439	974
Market Penetration	2007	0%	0%	0%	0%	1%	0%	0%	0%	0%	0%	0%	0%	0%
	2008	0%	0%	1%	0%	1%	0%	1%	0%	0%	0%	0%	0%	0%
	2009	0%	0%	2%	0%	2%	0%	1%	0%	0%	0%	0%	0%	0%
	2010	0%	0%	2%	0%	2%	0%	1%	0%	0%	0%	0%	2%	0%
	2011	0%	0%	2%	0%	3%	1%	1%	0%	0%	0%	0%	3%	0%
	2012	0%	0%	2%	1%	4%	1%	1%	0%	0%	0%	0%	3%	1%
	2013	0%	0%	2%	1%	5%	1%	2%	0%	0%	0%	0%	4%	1%
	2014	0%	0%	2%	2%	5%	3%	4%	0%	0%	0%	0%	6%	2%
	2015	0%	0%	3%	2%	5%	4%	4%	1%	0%	0%	0%	10%	3%

		Rhode Island	South Carolina	South Dakota	Tennessee	Texas	Utah	Vermont	Virginia	Washington	DC	West Virginia	Wisconsin	Wyoming
Cumulative installations	2007	1	2	1	1	6	1	2	1	6	1	1	6	1
	2008	1	2	1	9	9	1	2	1	9	1	1	11	1
	2009	1	2	1	9	17	1	2	1	11	1	1	11	1
	2010	1	2	1	9	41	1	3	1	17	4	1	13	1
	2011	2	3	1	9	52	1	3	1	23	4	1	18	1
	2012	3	6	1	9	80	1	5	2	32	7	1	25	1
	2013	5	10	1	9	108	1	6	4	52	25	1	35	1
	2014	8	21	1	14	199	4	9	23	106	25	2	57	1
	2015	10	27	1	27	350	7	13	38	106	61	5	71	1
Annual Installations	2007	1	1	0	0	1	1	1	1	3	1	0	2	0
	2008	0	0	0	8	3	0	0	1	3	0	0	5	0
	2009	0	0	0	0	8	0	0	0	2	0	0	0	0
	2010	1	0	0	0	25	0	1	0	6	3	0	2	0
	2011	1	1	0	0	11	0	1	0	6	0	0	5	0
	2012	1	3	0	0	28	0	1	1	9	2	0	7	0
	2013	2	4	0	0	28	0	1	2	19	18	0	9	0
	2014	3	11	0	5	91	4	3	19	55	0	1	22	0
	2015	2	6	0	13	151	3	4	14	0	37	4	14	0
Installers Required	2007	4	7	0	0	8	4	7	4	22	4	0	15	0
	2008	0	0	0	59	24	0	2	4	23	1	0	37	0
	2009	1	0	0	0	53	0	0	0	16	2	0	0	0
	2010	5	1	0	0	165	0	4	0	37	20	0	13	0
	2011	5	4	0	0	68	0	4	0	39	0	0	30	1
	2012	7	19	0	0	162	0	7	6	54	13	0	43	1
	2013	9	22	0	0	147	0	7	10	102	96	0	50	1
	2014	16	52	2	24	445	19	15	93	267	0	5	109	1
	2015	9	27	1	58	668	13	19	63	0	163	16	62	1
Market Penetration	2007	0%	0%	0%	0%	0%	0%	0%	0%	0%	0%	0%	0%	0%
	2008	0%	0%	0%	0%	0%	0%	0%	0%	0%	0%	0%	0%	0%
	2009	0%	0%	0%	0%	0%	0%	0%	0%	0%	0%	0%	0%	0%
	2010	0%	0%	0%	0%	0%	0%	0%	0%	0%	0%	0%	0%	0%
	2011	0%	0%	0%	0%	0%	0%	0%	0%	0%	0%	0%	0%	0%
	2012	0%	0%	0%	0%	0%	0%	0%	0%	0%	0%	0%	0%	0%
	2013	0%	0%	0%	0%	0%	0%	1%	0%	0%	1%	0%	0%	0%
	2014	1%	0%	0%	0%	0%	0%	1%	0%	1%	1%	0%	0%	0%
	2015	1%	0%	0%	0%	0%	0%	1%	0%	1%	3%	0%	1%	0%

Table A-15. State-by-State Results for the Focused Policy Case, SAI System Pricing

		Alabama	Alaska	Arizona	Arkansas	California	Colorado	Connecticut	Delaware	Florida	Georgia	Hawaii	Idaho	Illinois
Cumulative installations	2007	1	1	14	1	499	20	3	1	1	1	6	1	1
	2008	1	1	41	1	1,059	34	3	16	2	1	11	1	1
	2009	1	1	72	1	1,605	35	11	16	3	3	11	2	1
	2010	1	1	122	1	2,598	36	20	16	4	7	12	2	1
	2011	1	1	187	1	2,598	73	21	16	25	11	12	2	1
	2012	1	1	268	1	3,009	75	30	16	39	16	14	2	5
	2013	3	1	313	1	4,471	77	42	16	57	22	18	2	10
	2014	7	1	360	1	6,071	78	107	16	79	29	35	3	18
	2015	14	2	408	1	8,325	120	171	19	163	41	58	4	33
Annual Installations	2007	0	0	5	0	166	6	1	1	1	0	3	0	0
	2008	0	0	27	0	561	14	0	15	1	0	5	0	0
	2009	0	0	31	0	546	1	8	0	1	2	0	1	0
	2010	0	0	50	0	992	1	9	0	1	5	1	0	0
	2011	0	0	65	0	0	37	2	0	22	4	0	0	1
	2012	0	0	81	0	411	2	9	0	14	5	3	0	4
	2013	2	0	45	0	1,462	2	13	0	18	6	4	0	5
	2014	4	0	47	0	1,600	2	65	1	22	7	17	1	8
	2015	8	1	49	0	2,254	42	63	3	84	12	23	2	15
Installers Required	2007	0	0	35	0	1,232	48	7	4	4	0	23	0	0
	2008	0	0	171	0	4,037	102	2	107	6	0	35	1	0
	2009	0	0	214	0	3,798	5	54	0	6	15	0	6	0
	2010	0	0	339	0	6,663	5	62	0	6	32	5	0	0
	2011	0	0	407	0	0	235	10	0	135	24	0	0	6
	2012	0	0	467	0	2,384	9	49	0	82	28	15	0	22
	2013	12	0	239	0	7,801	9	67	9	96	32	22	0	28
	2014	20	0	228	0	7,799	8	316	41	105	34	83	5	37
	2015	34	5	214	0	9,951	184	280	12	369	51	101	7	65
Market Penetration	2007	0%	0%	0%	0%	1%	0%	0%	0%	0%	0%	0%	0%	0%
	2008	0%	0%	0%	0%	2%	0%	0%	1%	0%	0%	1%	0%	0%
	2009	0%	0%	1%	0%	3%	0%	0%	1%	0%	0%	1%	0%	0%
	2010	0%	0%	1%	0%	4%	0%	0%	1%	0%	0%	1%	0%	0%
	2011	0%	0%	1%	0%	4%	1%	0%	2%	0%	0%	0%	0%	0%
	2012	0%	0%	2%	0%	4%	1%	1%	3%	0%	0%	1%	0%	0%
	2013	0%	0%	2%	0%	6%	1%	1%	6%	0%	0%	1%	0%	0%
	2014	0%	0%	2%	0%	8%	1%	2%	8%	0%	0%	1%	0%	0%
	2015	0%	0%	2%	0%	10%	1%	3%	12%	0%	0%	2%	0%	0%

		Indiana	Iowa	Kansas	Kentucky	Louisiana	Maine	Maryland	Mass.	Michigan	Minnesota	Mississippi	Missouri
Cumulative Installations	2007	1	1	1	1	1	2	2	7	1	1	1	1
	2008	1	1	1	2	1	3	3	9	1	1	1	1
	2009	1	1	1	2	1	100	15	10	1	1	1	1
	2010	1	1	1	2	1	302	15	12	2	1	1	1
	2011	1	1	1	2	1	302	35	30	3	1	2	1
	2012	1	1	1	2	1	302	35	42	5	1	4	1
	2013	1	1	1	3	2	319	89	55	9	1	5	1
	2014	1	1	1	6	4	424	89	124	14	1	7	1
	2015	1	1	1	12	7	536	212	194	24	2	11	3
Annual Installations	2007	0	0	0	0	0	1	1	2	1	0	0	0
	2008	0	0	0	2	0	1	1	2	0	0	0	0
	2009	0	0	0	0	0	97	13	2	0	0	0	0
	2010	0	0	0	0	0	203	0	2	1	0	1	0
	2011	0	0	0	0	0	0	20	17	1	0	1	0
	2012	0	0	0	0	0	0	0	12	2	0	1	0
	2013	0	0	0	1	1	17	53	13	4	0	1	0
	2014	0	0	0	3	1	106	0	70	5	0	2	0
	2015	0	1	0	6	4	112	123	70	11	1	4	2
Installers Required	2007	0	0	0	0	0	7	7	16	4	0	0	0
	2008	0	0	0	12	0	5	4	13	0	2	0	0
	2009	0	0	0	0	0	674	89	13	0	0	0	0
	2010	0	0	0	0	0	1,360	0	12	6	0	6	0
	2011	0	0	0	0	0	0	125	108	6	0	6	0
	2012	0	0	0	0	3	0	0	70	13	0	7	0
	2013	0	0	0	6	6	88	284	69	20	0	8	0
	2014	0	0	0	13	7	515	0	341	25	0	9	0
	2015	0	3	0	28	17	494	545	308	47	5	19	11
Market Penetration	2007	0%	0%	0%	0%	0%	0%	0%	0%	0%	0%	0%	0%
	2008	0%	0%	0%	0%	0%	0%	0%	0%	0%	0%	0%	0%
	2009	0%	0%	0%	0%	0%	6%	0%	0%	0%	0%	0%	0%
	2010	0%	0%	0%	0%	0%	17%	0%	0%	0%	0%	0%	0%
	2011	0%	0%	0%	0%	0%	17%	0%	0%	0%	0%	0%	0%
	2012	0%	0%	0%	0%	0%	16%	0%	0%	0%	0%	0%	0%
	2013	0%	0%	0%	0%	0%	16%	1%	1%	0%	0%	0%	0%
	2014	0%	0%	0%	0%	0%	20%	1%	1%	0%	0%	0%	0%
	2015	0%	0%	0%	0%	0%	24%	2%	2%	0%	0%	0%	0%

Table A-15 (Continued)

		Montana	Nebraska	Nevada	NH	New Jersey	New Mexico	New York	North Carolina	North Dakota	Ohio	Oklahoma	Oregon	Pennsylvania
Cumulative installations	2007	1	1	15	1	69	9	32	3	1	2	1	2	9
	2008	1	1	79	1	103	12	128	6	1	5	1	48	9
	2009	1	1	107	2	140	14	134	29	1	5	1	48	23
	2010	1	1	109	9	322	15	140	140	1	7	1	48	45
	2011	1	1	140	9	322	16	146	140	1	9	2	48	58
	2012	1	1	143	21	472	26	153	140	1	12	3	48	96
	2013	2	1	175	30	405	43	159	140	1	16	4	48	166
	2014	2	1	179	33	502	79	275	140	1	20	5	48	290
	2015	5	2	314	33	614	123	435	154	1	34	7	48	442
Annual installations	2007	0	0	7	1	29	3	10	1	0	1	0	1	3
	2008	0	0	64	0	34	4	95	3	0	3	0	46	0
	2009	0	0	28	1	37	2	6	23	0	0	0	0	13
	2010	0	0	2	7	183	1	6	112	0	1	1	0	23
	2011	0	0	30	0	0	1	6	0	0	2	1	0	13
	2012	0	0	3	12	150	10	6	0	0	3	1	0	38
	2013	0	0	32	9	-68	17	6	0	0	4	1	0	70
	2014	1	0	4	2	98	36	116	0	0	4	1	0	124
	2015	3	2	136	0	111	44	160	13	0	14	2	0	152
Installers Required	2007	0	0	56	4	213	24	71	7	0	4	0	4	19
	2008	3	0	12	0	243	27	43	22	0	25	0	329	1
	2009	0	0	198	8	255	12	42	158	0	0	0	0	93
	2010	0	0	16	48	1,226	4	41	750	0	9	6	0	152
	2011	0	0	189	0	0	7	39	0	0	15	3	0	78
	2012	1	0	17	68	868	56	37	0	0	18	4	0	221
	2013	3	0	172	50	-360	92	34	0	0	19	5	0	373
	2014	4	1	18	11	475	176	566	0	0	20	5	0	606
	2015	13	8	599	2	492	196	705	58	0	63	11	0	671
Market Penetration	2007	0%	0%	0%	0%	1%	0%	0%	0%	0%	0%	0%	0%	0%
	2008	0%	0%	1%	0%	1%	0%	1%	0%	0%	0%	0%	1%	0%
	2009	0%	0%	2%	0%	2%	0%	1%	0%	0%	0%	0%	1%	0%
	2010	0%	0%	2%	1%	4%	0%	1%	1%	0%	0%	0%	1%	0%
	2011	0%	0%	2%	1%	3%	0%	1%	1%	0%	0%	0%	1%	0%
	2012	0%	0%	2%	1%	5%	1%	1%	1%	0%	0%	0%	1%	1%
	2013	0%	0%	2%	2%	4%	1%	1%	1%	0%	0%	0%	1%	1%
	2014	0%	0%	2%	2%	5%	2%	1%	1%	0%	0%	0%	1%	2%
	2015	0%	0%	3%	2%	5%	3%	2%	1%	0%	0%	0%	1%	3%

		Rhode Island	South Carolina	South Dakota	Tennessee	Texas	Utah	Vermont	Virginia	Washington	DC	West Virginia	Wisconsin	Wyoming
Cumulative Installations	2007	1	2	1	1	6	1	2	1	6	1	1	6	1
	2008	1	2	1	11	9	1	2	1	10	1	1	12	1
	2009	1	2	1	11	13	1	2	1	19	1	1	12	1
	2010	3	2	1	11	21	1	3	1	36	13	1	20	1
	2011	3	2	1	11	27	1	3	2	37	13	1	20	1
	2012	4	3	1	11	35	1	4	4	53	13	1	25	1
	2013	6	4	1	16	45	1	6	8	109	24	1	35	2
	2014	8	6	1	24	56	1	7	13	211	24	2	48	2
	2015	12	11	1	39	83	1	10	22	211	37	4	70	3
Annual Installations	2007	1	1	0	0	1	1	1	1	3	1	0	2	0
	2008	0	0	0	11	3	0	0	1	4	0	0	6	0
	2009	0	0	0	0	3	0	0	0	10	0	0	0	0
	2010	2	0	0	0	8	0	1	0	16	11	0	8	0
	2011	0	0	0	0	6	0	0	0	1	0	0	0	0
	2012	1	1	0	0	8	0	1	3	16	0	0	5	0
	2013	2	1	0	5	9	0	1	4	55	11	0	10	0
	2014	3	2	0	8	11	0	2	5	103	0	1	13	0
	2015	4	5	0	15	27	1	2	10	0	13	2	21	1
Installers Required	2007	4	7	0	0	8	4	7	4	22	4	0	15	0
	2008	0	0	0	77	25	0	1	4	27	1	0	43	0
	2009	2	0	0	0	24	0	0	0	67	2	0	0	0
	2010	12	0	0	0	57	0	7	0	109	77	0	52	1
	2011	0	1	0	0	37	0	0	3	6	0	0	0	1
	2012	7	4	0	0	47	0	5	15	95	0	0	29	2
	2013	11	7	0	27	49	0	8	19	296	93	2	56	2
	2014	12	10	0	40	55	0	9	23	501	0	5	65	2
	2015	17	21	0	66	119	3	10	43	0	56	9	94	3
Market Penetration	2007	0%	0%	0%	0%	0%	0%	0%	0%	0%	0%	0%	0%	0%
	2008	0%	0%	0%	0%	0%	0%	0%	0%	0%	0%	0%	0%	0%
	2009	0%	0%	0%	0%	0%	0%	0%	0%	0%	0%	0%	0%	0%
	2010	0%	0%	0%	0%	0%	0%	0%	0%	0%	1%	0%	0%	0%
	2011	0%	0%	0%	0%	0%	0%	0%	0%	0%	1%	0%	0%	0%
	2012	0%	0%	0%	0%	0%	0%	0%	0%	0%	1%	0%	0%	0%
	2013	0%	0%	0%	0%	0%	0%	1%	0%	1%	1%	0%	0%	0%
	2014	1%	0%	0%	0%	0%	0%	1%	0%	2%	1%	0%	0%	0%
	2015	1%	0%	0%	0%	0%	0%	1%	0%	1%	2%	0%	1%	0%

Table A-16. State-by-State Results for the Best Case, BAU System Pricing

		Alabama	Alaska	Arizona	Arkansas	California	Colorado	Connecticut	Delaware	Florida	Georgia	Hawaii	Idaho	Illinois
Cumulative Installations	2007	1	1	14	1	499	20	3	1	1	1	6	1	1
	2008	1	1	41	1	1,215	34	3	17	2	1	13	1	1
	2009	1	1	72	1	1,302	35	9	17	3	1	13	1	1
	2010	1	1	122	1	1,302	36	11	17	4	1	13	1	1
	2011	1	1	187	1	1,971	73	16	17	8	1	18	1	1
	2012	1	1	268	1	2,940	75	24	17	16	1	24	1	1
	2013	2	1	313	1	3,429	77	32	17	21	2	31	1	1
	2014	11	1	360	1	5,293	78	49	17	36	11	40	1	1
	2015	27	2	408	1	6,917	120	57	18	72	15	54	1	3
Annual Installations	2007	0	0	5	0	186	8	1	1	1	0	3	0	0
	2008	0	0	27	0	716	14	0	16	1	0	7	0	0
	2009	0	0	31	0	88	1	6	0	1	0	0	0	0
	2010	0	0	50	0	0	1	2	0	1	0	0	0	0
	2011	0	0	65	0	668	37	5	0	4	0	5	0	0
	2012	0	0	81	0	970	2	8	0	8	0	6	0	0
	2013	1	0	45	0	489	2	8	0	5	1	7	0	0
	2014	9	0	47	0	1,864	2	17	0	15	9	9	0	0
	2015	16	1	49	0	1,624	42	8	1	37	3	14	0	3
Installers Required	2007	0	0	35	0	1,232	48	7	4	4	0	23	0	0
	2008	0	0	171	0	5,156	102	2	116	6	0	53	1	0
	2009	0	0	214	0	609	5	41	0	6	0	0	3	0
	2010	0	0	339	0	0	5	16	0	6	0	0	0	0
	2011	0	0	407	0	4,180	235	28	0	25	0	30	0	0
	2012	0	0	467	0	5,621	9	47	0	46	0	36	0	0
	2013	6	2	239	0	2,609	9	44	1	29	7	36	0	0
	2014	44	2	228	0	9,086	8	81	48	71	46	44	0	0
	2015	71	3	214	0	7,171	184	37	6	163	15	63	0	12
Market Penetration	2007	0%	0%	0%	0%	1%	0%	0%	0%	0%	0%	0%	0%	0%
	2008	0%	0%	0%	0%	2%	0%	0%	1%	0%	0%	1%	0%	0%
	2009	0%	0%	1%	0%	2%	0%	0%	1%	0%	0%	1%	0%	0%
	2010	0%	0%	1%	0%	2%	0%	0%	1%	0%	0%	1%	0%	0%
	2011	0%	0%	1%	0%	3%	1%	0%	1%	0%	0%	1%	0%	0%
	2012	0%	0%	2%	0%	4%	1%	0%	1%	0%	0%	1%	0%	0%
	2013	0%	0%	2%	0%	5%	1%	1%	1%	0%	0%	1%	0%	0%
	2014	0%	0%	2%	0%	7%	1%	1%	1%	0%	0%	1%	0%	0%
	2015	0%	0%	2%	0%	9%	1%	1%	1%	0%	0%	2%	0%	0%

		Indiana	Iowa	Kansas	Kentucky	Louisiana	Maine	Maryland	Mass.	Michigan	Minnesota	Mississippi	Missouri
Cumulative Installations	2007	1	1	1	1	1	2	2	7	1	1	1	1
	2008	1	1	1	2	1	3	3	9	1	1	1	1
	2009	1	1	1	2	1	59	15	10	1	1	1	1
	2010	1	1	1	2	1	195	15	12	1	1	1	1
	2011	1	1	1	2	1	195	35	29	1	1	1	1
	2012	1	1	1	2	1	208	35	43	1	1	5	1
	2013	1	1	1	2	1	268	89	57	1	1	9	1
	2014	1	1	1	2	1	344	89	84	2	1	19	1
	2015	1	1	1	3	1	417	212	97	3	1	27	1
Annual Installations	2007	0	0	0	0	0	1	1	2	1	0	0	0
	2008	0	0	0	2	0	1	1	2	0	0	0	0
	2009	0	0	0	0	0	56	13	2	0	0	0	0
	2010	0	0	0	0	0	137	0	2	0	0	0	0
	2011	0	0	0	0	0	0	20	17	0	0	0	0
	2012	0	0	0	0	0	12	0	14	0	0	5	0
	2013	0	0	0	0	0	61	53	14	0	0	4	0
	2014	0	0	0	0	0	76	0	27	1	0	10	0
	2015	0	0	0	1	0	73	123	13	1	0	7	0
Installers Required	2007	0	0	0	0	0	7	7	16	4	0	0	0
	2008	0	0	0	14	0	4	4	13	0	2	0	0
	2009	0	0	0	0	0	389	89	13	0	0	0	0
	2010	0	0	0	0	0	919	0	12	0	0	0	0
	2011	0	0	0	0	0	0	125	107	0	0	0	0
	2012	0	0	0	0	0	71	0	82	0	0	28	0
	2013	0	0	0	0	0	323	284	75	0	0	20	0
	2014	0	0	0	0	0	369	0	130	6	0	49	0
	2015	0	0	0	4	0	321	545	58	4	0	33	0
Market Penetration	2007	0%	0%	0%	0%	0%	0%	0%	0%	0%	0%	0%	0%
	2008	0%	0%	0%	0%	0%	0%	0%	0%	0%	0%	0%	0%
	2009	0%	0%	0%	0%	0%	4%	0%	0%	0%	0%	0%	0%
	2010	0%	0%	0%	0%	0%	11%	0%	0%	0%	0%	0%	0%
	2011	0%	0%	0%	0%	0%	11%	0%	0%	0%	0%	0%	0%
	2012	0%	0%	0%	0%	0%	11%	0%	0%	0%	0%	0%	0%
	2013	0%	0%	0%	0%	0%	13%	1%	1%	0%	0%	0%	0%
	2014	0%	0%	0%	0%	0%	16%	1%	1%	0%	0%	0%	0%
	2015	0%	0%	0%	0%	0%	19%	2%	1%	0%	0%	0%	0%

Table A-16 (Continued)

		Montana	Nebraska	Nevada	NH	New Jersey	New Mexico	New York	North Carolina	North Dakota	Ohio	Oklahoma	Oregon	Pennsylvania
Cumulative installations	2007	1	1	15	1	69	9	32	3	1	2	1	2	9
	2008	1	1	79	1	103	12	128	6	1	6	1	6	10
	2009	1	1	107	1	140	14	134	20	1	6	1	9	23
	2010	1	1	109	6	194	14	140	51	1	6	1	101	37
	2011	1	1	140	8	253	15	146	51	1	6	1	149	58
	2012	1	1	143	16	346	21	153	76	1	6	1	202	96
	2013	1	1	175	22	472	26	159	76	1	7	1	264	166
	2014	1	1	179	33	518	37	303	76	1	11	1	360	290
	2015	2	1	203	33	614	55	357	154	1	19	2	468	343
Annual Installations	2007	0	0	7	1	29	3	10	1	0	1	0	1	3
	2008	0	0	64	0	34	4	95	3	0	4	0	4	1
	2009	0	0	28	0	37	1	6	14	0	0	0	3	13
	2010	0	0	2	4	55	0	6	31	0	0	0	92	14
	2011	0	0	30	3	59	1	6	0	0	0	0	48	21
	2012	0	0	3	8	93	6	6	25	0	0	0	53	38
	2013	0	0	32	6	126	5	6	0	0	1	0	62	70
	2014	0	0	4	11	46	11	144	0	0	4	1	96	124
	2015	1	0	24	0	96	19	55	77	0	8	0	108	53
Installers Required	2007	0	0	56	4	213	24	71	7	0	4	0	4	19
	2008	3	0	12	0	243	26	43	22	0	25	0	30	5
	2009	0	0	198	3	255	9	42	98	0	0	0	18	89
	2010	0	0	16	28	366	0	41	205	0	0	0	616	93
	2011	0	0	189	17	366	7	39	0	0	0	0	302	133
	2012	0	0	17	44	537	36	37	144	0	0	0	307	221
	2013	0	0	172	30	672	28	34	2	0	6	0	332	373
	2014	0	0	18	54	226	52	700	2	0	21	4	469	606
	2015	4	0	107	2	423	82	241	341	0	33	2	476	233
Market Penetration	2007	0%	0%	0%	0%	1%	0%	0%	0%	0%	0%	0%	0%	0%
	2008	0%	0%	1%	0%	1%	0%	1%	0%	0%	0%	0%	0%	0%
	2009	0%	0%	2%	0%	2%	0%	1%	0%	0%	0%	0%	0%	0%
	2010	0%	0%	2%	0%	2%	0%	1%	0%	0%	0%	0%	2%	0%
	2011	0%	0%	2%	0%	3%	0%	1%	0%	0%	0%	0%	2%	0%
	2012	0%	0%	2%	1%	3%	1%	1%	0%	0%	0%	0%	3%	1%
	2013	0%	0%	2%	1%	4%	1%	1%	0%	0%	0%	0%	3%	1%
	2014	0%	0%	2%	2%	5%	1%	2%	0%	0%	0%	0%	4%	2%
	2015	0%	0%	2%	2%	5%	1%	2%	1%	0%	0%	0%	6%	2%

		Rhode Island	South Carolina	South Dakota	Tennessee	Texas	Utah	Vermont	Virginia	Washington	DC	West Virginia	Wisconsin	Wyoming
Cumulative Installations	2007	1	2	1	1	6	1	2	1	6	1	1	6	1
	2008	1	2	1	12	9	1	2	1	10	1	1	11	1
	2009	1	2	1	12	18	1	2	1	15	1	1	11	1
	2010	1	2	1	12	34	1	2	1	20	5	1	11	1
	2011	1	2	1	12	39	1	3	1	28	5	1	12	1
	2012	2	2	1	12	50	1	4	1	41	7	1	19	1
	2013	3	2	1	12	58	1	4	1	54	22	1	26	1
	2014	5	10	1	17	108	1	6	2	75	22	1	42	1
	2015	6	12	1	28	195	1	7	5	75	22	1	47	1
Annual Installations	2007	1	1	0	0	1	1	1	1	3	1	0	2	0
	2008	0	0	0	11	3	0	0	1	4	0	0	5	0
	2009	0	0	0	0	9	0	0	0	5	0	0	0	0
	2010	0	0	0	0	16	0	0	0	5	3	0	0	0
	2011	0	0	0	0	5	0	0	0	8	0	0	1	0
	2012	1	0	0	0	11	0	1	0	13	2	0	7	0
	2013	1	0	0	0	8	0	1	0	13	16	0	7	0
	2014	2	8	0	5	51	0	2	1	21	0	0	16	0
	2015	1	1	0	11	87	0	1	3	0	0	0	5	0
Installers Required	2007	4	7	0	0	8	4	7	4	22	4	0	15	0
	2008	0	0	0	79	25	0	1	4	29	1	0	33	0
	2009	0	0	0	0	63	0	0	0	36	2	0	0	0
	2010	0	0	0	0	105	0	1	0	31	23	0	0	0
	2011	3	0	0	0	30	0	2	0	49	0	0	7	0
	2012	7	0	0	0	63	0	5	0	77	10	0	41	0
	2013	5	0	0	0	44	0	4	0	71	84	0	38	0
	2014	10	40	0	27	247	0	8	4	104	0	0	80	0
	2015	4	6	0	47	382	0	3	13	0	0	0	23	0
Market Penetration	2007	0%	0%	0%	0%	0%	0%	0%	0%	0%	0%	0%	0%	0%
	2008	0%	0%	0%	0%	0%	0%	0%	0%	0%	0%	0%	0%	0%
	2009	0%	0%	0%	0%	0%	0%	0%	0%	0%	0%	0%	0%	0%
	2010	0%	0%	0%	0%	0%	0%	0%	0%	0%	0%	0%	0%	0%
	2011	0%	0%	0%	0%	0%	0%	0%	0%	0%	0%	0%	0%	0%
	2012	0%	0%	0%	0%	0%	0%	0%	0%	0%	0%	0%	0%	0%
	2013	0%	0%	0%	0%	0%	0%	0%	0%	0%	1%	0%	0%	0%
	2014	0%	0%	0%	0%	0%	0%	1%	0%	1%	1%	0%	0%	0%
	2015	0%	0%	0%	0%	0%	0%	1%	0%	1%	1%	0%	0%	0%

Table A-17. State-by-State Results for the Best Case, SAI System Pricing

	Year	Alabama	Alaska	Arizona	Arkansas	California	Colorado	Connecticut	Delaware	Florida	Georgia	Hawaii	Idaho	Illinois
Cumulative Installations	2007	1	1	14	1	499	20	3	1	1	1	6	1	1
	2008	1	1	41	1	1,416	34	3	20	2	1	15	1	1
	2009	1	1	72	1	1,748	35	11	20	3	1	15	1	1
	2010	1	1	122	1	2,135	36	17	20	4	4	23	1	1
	2011	4	1	187	1	3,233	73	28	20	24	8	33	1	1
	2012	10	2	268	1	4,997	75	43	20	51	12	47	1	1
	2013	23	3	313	1	6,571	77	82	20	104	18	90	1	1
	2014	54	3	382	1	10,449	119	291	44	224	32	155	2	13
	2015	87	5	843	1	14,133	232	409	70	330	41	243	3	36
Annual Installations	2007	0	0	5	0	166	6	1	1	1	0	3	0	0
	2008	0	0	27	0	917	14	0	19	1	0	9	0	0
	2009	0	0	31	0	332	1	8	0	1	1	0	1	0
	2010	1	0	50	0	387	1	6	0	1	3	7	0	0
	2011	2	1	65	0	1,098	37	11	0	20	3	10	0	0
	2012	6	0	81	0	1,764	2	15	0	27	5	14	0	0
	2013	13	1	45	0	1,574	2	40	0	53	6	43	0	1
	2014	31	1	69	0	3,878	42	209	25	120	14	66	0	12
	2015	33	2	461	1	3,684	112	117	25	106	9	88	2	23
Installers Required	2007	0	0	35	0	1,232	48	7	4	4	0	23	0	0
	2008	0	0	171	0	6,602	102	2	136	6	0	68	1	0
	2009	0	0	214	0	2,313	5	56	0	6	5	0	5	0
	2010	6	1	339	0	2,599	5	42	0	6	21	49	0	0
	2011	14	4	407	0	6,868	235	66	0	128	21	65	0	0
	2012	36	3	467	0	10,222	9	86	0	154	26	79	0	0
	2013	70	4	239	0	8,399	12	211	0	282	31	228	0	4
	2014	150	4	335	0	18,904	205	1,020	120	586	68	319	2	57
	2015	146	8	2,036	2	16,264	496	518	111	469	38	388	8	101
Market Penetration	2007	0%	0%	0%	0%	1%	0%	0%	0%	0%	0%	0%	0%	0%
	2008	0%	0%	0%	0%	3%	0%	0%	2%	0%	0%	1%	0%	0%
	2009	0%	0%	1%	0%	3%	0%	0%	1%	0%	0%	1%	0%	0%
	2010	0%	0%	1%	0%	3%	0%	0%	1%	0%	0%	1%	0%	0%
	2011	0%	0%	1%	0%	5%	1%	1%	1%	0%	0%	1%	0%	0%
	2012	0%	0%	2%	0%	7%	1%	1%	1%	0%	0%	2%	0%	0%
	2013	0%	0%	2%	0%	9%	1%	2%	1%	0%	0%	3%	0%	0%
	2014	0%	0%	2%	0%	14%	1%	5%	2%	0%	0%	6%	0%	0%
	2015	1%	0%	4%	0%	17%	2%	7%	3%	1%	0%	8%	0%	0%

		Indiana	Iowa	Kansas	Kentucky	Louisiana	Maine	Maryland	Mass.	Michigan	Minnesota	Mississippi	Missouri
Cumulative Installations	2007	1	1	1	1	1	2	2	7	1	1	1	1
	2008	1	1	1	3	2	3	3	9	1	1	1	1
	2009	1	1	1	3	2	100	15	10	1	1	1	1
	2010	1	1	1	3	2	295	15	12	1	1	1	1
	2011	1	1	1	3	2	295	35	47	1	1	5	1
	2012	1	1	1	3	2	313	35	99	3	1	14	1
	2013	1	1	1	3	3	411	89	174	8	3	22	1
	2014	1	1	1	18	28	529	101	456	36	16	48	1
	2015	1	1	1	35	34	661	212	659	52	22	92	3
Annual Installations	2007	0	0	0	0	0	1	1	2	1	0	0	0
	2008	0	0	0	2	2	1	1	2	0	0	0	0
	2009	0	0	0	0	0	97	13	2	0	0	0	0
	2010	0	0	0	0	0	196	0	2	0	0	0	0
	2011	0	0	0	0	0	0	20	35	0	0	5	0
	2012	0	0	0	0	0	18	0	51	2	1	9	0
	2013	0	0	0	0	0	97	53	75	5	2	7	0
	2014	0	0	0	15	26	119	12	283	28	13	26	0
	2015	0	1	0	16	6	132	111	202	16	6	45	2
Installers Required	2007	0	0	0	0	0	7	7	16	4	0	0	0
	2008	0	0	0	17	14	7	4	13	0	2	0	0
	2009	0	0	0	0	0	672	89	13	0	0	0	0
	2010	0	0	0	0	0	1,314	0	12	0	0	2	0
	2011	0	0	0	0	0	0	125	220	2	0	29	0
	2012	0	0	0	0	0	104	0	298	9	4	52	0
	2013	0	0	0	2	2	520	284	398	27	9	39	0
	2014	0	1	0	74	125	578	61	1,379	138	63	126	0
	2015	0	3	0	73	26	581	489	893	71	25	197	11
Market Penetration	2007	0%	0%	0%	0%	0%	0%	0%	0%	0%	0%	0%	0%
	2008	0%	0%	0%	0%	0%	0%	0%	0%	0%	0%	0%	0%
	2009	0%	0%	0%	0%	0%	6%	0%	0%	0%	0%	0%	0%
	2010	0%	0%	0%	0%	0%	17%	0%	0%	0%	0%	0%	0%
	2011	0%	0%	0%	0%	0%	16%	0%	1%	0%	0%	0%	0%
	2012	0%	0%	0%	0%	0%	16%	0%	1%	0%	0%	0%	0%
	2013	0%	0%	0%	0%	0%	20%	1%	2%	0%	0%	0%	0%
	2014	0%	0%	0%	0%	0%	25%	1%	5%	0%	0%	1%	0%
	2015	0%	0%	0%	0%	0%	30%	2%	7%	0%	0%	1%	0%

Table A-17 (Continued)

		Montana	Nebraska	Nevada	NH	New Jersey	New Mexico	New York	North Carolina	North Dakota	Ohio	Oklahoma	Oregon	Pennsylvania
Cumulative Installations	2007	1	1	15	1	69	9	32	3	1	2	1	2	9
	2008	1	1	79	1	103	14	128	6	1	7	1	6	10
	2009	1	1	107	2	140	16	134	27	1	7	1	9	23
	2010	1	1	109	9	316	16	140	134	1	7	1	157	41
	2011	1	1	140	13	316	23	146	134	1	8	1	229	61
	2012	1	1	143	22	321	32	265	134	1	11	2	325	115
	2013	2	1	175	22	449	61	482	134	1	15	3	425	210
	2014	5	4	373	33	502	159	913	134	1	29	5	672	479
	2015	8	5	668	80	735	259	1,098	154	1	66	7	1,106	756
Annual Installations	2007	0	0	7	1	29	3	10	1	0	1	0	1	3
	2008	0	0	64	0	34	5	95	3	0	5	0	4	1
	2009	0	0	28	1	37	2	6	21	0	0	0	3	12
	2010	0	0	2	7	176	0	6	107	0	0	0	148	18
	2011	0	0	30	4	0	7	6	0	0	1	0	73	20
	2012	0	0	3	9	6	9	118	0	0	3	1	95	54
	2013	1	0	32	0	128	29	218	0	0	5	1	100	95
	2014	3	3	198	11	53	98	431	0	0	13	3	247	268
	2015	3	1	295	47	233	100	185	20	0	37	1	435	277
Installers Required	2007	0	0	56	4	213	24	71	7	0	4	0	4	19
	2008	3	0	12	0	243	35	43	23	0	33	0	30	8
	2009	0	0	198	8	255	14	42	145	0	0	0	18	86
	2010	0	0	16	46	1,182	0	41	718	0	3	1	994	120
	2011	0	0	189	26	0	43	39	0	0	8	2	454	125
	2012	0	0	17	50	89	55	687	0	0	15	4	553	316
	2013	6	1	172	0	683	154	1,161	0	0	25	4	535	508
	2014	17	13	966	53	257	479	2,100	0	0	65	14	1,203	1,308
	2015	13	6	1,304	209	1,028	440	815	88	1	163	6	1,919	1,223
Market Penetration	2007	0%	0%	0%	0%	1%	0%	0%	0%	0%	0%	0%	0%	0%
	2008	0%	0%	1%	0%	1%	0%	1%	0%	0%	0%	0%	0%	0%
	2009	0%	0%	2%	0%	2%	0%	1%	0%	0%	0%	0%	0%	0%
	2010	0%	0%	2%	1%	3%	0%	1%	1%	0%	0%	0%	2%	0%
	2011	0%	0%	2%	1%	3%	1%	1%	1%	0%	0%	0%	3%	0%
	2012	0%	0%	2%	1%	3%	1%	1%	1%	0%	0%	0%	5%	1%
	2013	0%	0%	2%	1%	4%	1%	3%	1%	0%	0%	0%	6%	1%
	2014	0%	0%	4%	2%	5%	4%	5%	1%	0%	0%	0%	8%	3%
	2015	0%	0%	7%	4%	6%	6%	5%	1%	0%	0%	0%	13%	5%

		Rhode Island	South Carolina	South Dakota	Tennessee	Texas	Utah	Vermont	Virginia	Washington	DC	West Virginia	Wisconsin	Wyoming
Cumulative Installations	2007	1	2	1	1	6	1	2	1	6	1	1	6	1
	2008	1	2	1	13	11	1	2	1	11	1	1	12	1
	2009	1	2	1	13	28	1	2	1	21	1	1	12	1
	2010	2	2	1	13	66	1	3	1	35	9	1	18	1
	2011	3	3	1	13	85	1	4	1	51	9	1	26	1
	2012	5	9	1	14	127	1	5	3	77	9	1	38	1
	2013	7	16	1	23	168	1	7	6	127	60	1	54	1
	2014	11	33	2	50	308	10	11	38	259	60	2	91	1
	2015	19	43	2	84	877	17	17	62	259	60	8	114	1
Annual Installations	2007	1	1	0	0	1	1	1	1	3	1	0	2	0
	2008	0	0	0	13	5	0	0	1	5	0	0	6	0
	2009	0	0	0	0	17	0	0	0	9	0	0	0	0
	2010	1	0	0	0	38	0	1	0	14	8	0	6	0
	2011	1	1	0	0	19	0	1	0	16	0	0	8	0
	2012	2	6	0	1	42	0	2	2	26	0	0	12	0
	2013	2	7	0	9	42	0	2	3	50	50	0	16	0
	2014	4	18	1	27	140	10	4	32	132	0	2	37	0
	2015	8	10	1	34	569	7	6	24	0	0	6	23	1
Installers Required	2007	4	7	0	0	8	4	7	4	22	4	0	15	0
	2008	0	0	0	93	38	0	2	4	38	1	0	43	0
	2009	0	0	0	0	117	0	0	0	65	2	0	0	0
	2010	7	2	0	0	254	0	5	0	96	54	0	39	0
	2011	7	6	0	0	119	0	6	0	99	0	0	50	0
	2012	10	32	0	4	242	0	9	12	152	0	0	72	0
	2013	13	37	0	46	223	0	9	16	265	284	1	83	0
	2014	22	86	5	131	682	47	20	155	646	0	8	182	1
	2015	34	44	3	152	2,512	32	26	105	0	0	26	103	3
Market Penetration	2007	0%	0%	0%	0%	0%	0%	0%	0%	0%	0%	0%	0%	0%
	2008	0%	0%	0%	0%	0%	0%	0%	0%	0%	0%	0%	0%	0%
	2009	0%	0%	0%	0%	0%	0%	0%	0%	0%	0%	0%	0%	0%
	2010	0%	0%	0%	0%	0%	0%	0%	0%	0%	1%	0%	0%	0%
	2011	0%	0%	0%	0%	0%	0%	0%	0%	0%	1%	0%	0%	0%
	2012	0%	0%	0%	0%	0%	0%	1%	0%	1%	1%	0%	0%	0%
	2013	1%	0%	0%	0%	0%	0%	1%	0%	1%	4%	0%	0%	0%
	2014	1%	0%	0%	0%	0%	0%	1%	0%	2%	3%	0%	1%	0%
	2015	1%	0%	0%	0%	1%	0%	2%	0%	2%	3%	0%	1%	0%

A-6. Input Data

Table A-18. Utilities Analyzed

State	Utility Name
AL	Alabama Power Co.
AK	Chugach
AZ	Arizona Public Service
AZ	Salt River Project
AZ	Tucson Electric Power
AK	Entergy Arkansas
CA	Southern California Edison
CA	Sacramento Municipal Utility District
CA	Pacific Gas and Electric Company
CA	San Diego Gas & Electric Company
CA	Los Angeles Department of Water and Power
CO	Public Service Company of Colorado
CO	Colorado Springs
CT	Connecticut Light and Power
DE	Conective (Delmarva Power)
FL	Florida Power & Light Co.
FL	Progress Energy Florida Inc
FL	Tampa Electric Company
GA	Georgia Power
HI	Hawaiian Electric Company (Oahu)
HI	Maui Electric Company
ID	Idaho Power
IL	Commonwealth Edison Co.
IL	Illinois Power Company
IN	PSI Energy Inc.
IA	IES Utilities (Mid America)
IA	Interstate Power and Light
KS	Kansas Gas & Electric Co
KS	Westar Energy Inc
KY	Kentucky Utilities Co
KY	Louisville Gas & Electric Co
KY	Kenergy Corporation
LA	Entergy (Louisiana Power & Light)
ME	Central Maine Power

State	Utility Name
ME	Bangor Hydro Electric Company
MD	BGE (Baltimore Gas and Electric)
MD	Potomac Electric Power Company
MA	NSTAR (Boston Edison)
MA	Massachusetts Electric Company
MI	Detroit Edison
MI	Consumers Energy Company
MN	Xcel Energy (Northern States Power)
MS	Entergy Mississippi (Mississippi Power and Light)
MS	Mississippi Power Company
MO	AmerenUE - Missouri (Union Electric)
MT	Northwestern Energy (Montana Power Company)
NE	Omaha Public Power District
NV	Nevada Power
NV	Sierra Pacific Power Company
NH	Public Service of New Hampshire
NH	Unitil Energy Systems
NJ	PSE&G (Public Service Electric and Gas Co.)
NJ	Jersey Central Power and Light Co.
NJ	Atlantic City Electrical Company
NM	PNM (Public Service Company of New Mexico)
NM	Southwest Public Service Company
NY	Niagara Mohawk
NY	New York State Electric and Gas Corp
NY	Consolidated Edison
NY	Long Island Power Authority
NC	Duke Power
NC	Progress Energy Carolinas Inc
ND	Northern States Power Co
OH	Ohio Power Company
OH	Ohio Edison
OH	Cincinnati Gas & Electric Company
OK	AEP (Public Service Company of Oklahoma)
OK	Oklahoma Gas and Electric Company
OR	PacifiCorp (Pacific Power)
OR	Portland General Electric Company
PA	PPL Electric Utilities
PA	PECO Energy Co

Table A-18. (Continued).

State	Utility Name
PA	West Penn Power Co.
RI	Narragansett Electric
SC	South Carolina Electric and Gas
SC	Duke Energy Corporation
SD	Xcel Energy (Northern States Power)
TN	Nashville Electric Service
TN	Knoxville Electric Board
TN	City of Memphis
TX	TXU Electric
TX	Reliant Energy Services
TX	Entergy Gulf States Inc
TX	Constellation New Energy Inc
TX	City of San Antonio
UT	PacifiCorp (Utah Power & Light)
VT	Green Mountain Power
VT	Central Vermont Public Service Corporation
VA	Dominion (Virginia Electric and Power)
VA	Appalachian Power Co
WA	Puget Sound Energy
WA	Snohomish County PUD No 1
WA	City of Seattle
DC	PEPCO
WV	American Electric (Appalachian Power)
WI	We Energies (Wisconsin Electric)
WI	Wisconsin Public Service Corporation
WY	PacifiCorp (Pacific Power)

Table A-19. IREC's Interconnection Assessments

State	Utility	Interconnection Policy Assessment
Alabama	Alabama Power Co.	Barrier
Alaska	Chugach	Good
Arizona	Arizona Public Service	Good
Arizona	Salt River Project	Good
Arizona	Tucson Electric Power	Good
Arkansas	Entergy Arkansas	Poor

State	Utility	Interconnection Policy Assessment
California	Southern California Edison	Fair
California	Sacramento Municipal Utility District	Fair
California	Pacific Gas and Electric Company	Fair
California	San Diego Gas & Electric Company	Fair
California	Los Angeles Department of Water and Power	Fair
Colorado	Public Service Company of Colorado	Fair
Colorado	Colorado Springs	Fair
Connecticut	Connecticut Light and Power	Poor
Delaware	Conective (Delmarva Power)	Barrier
Florida	Florida Power & Light Co.	Poor
Florida	Progress Energy Florida Inc	Poor
Florida	Tampa Electric Company	Poor
Georgia	Georgia Power	Fair
Hawaii	Hawaiian Electric Company (Oahu)	Barrier
Hawaii	Maui Electric Company	Barrier
Idaho	Idaho Power	Barrier
Illinois	Commonwealth Edison Co.	Barrier
Illinois	Illinois Power Company	Barrier
Indiana	PSI Energy Inc.	Poor
Iowa	IES Utilities (Mid American)	Poor
Iowa	Interstate Power and Light	Poor
Kansas	Kansas Gas & Electric Co	Barrier
Kansas	Westar Energy Inc	Barrier
Kentucky	Kentucky Utilities Co	Barrier
Kentucky	Louisville Gas & Electric Co	Barrier
Kentucky	Kenergy Corporation	Barrier
Louisiana	Entergy (Louisiana Power & Light)	Barrier
Maine	Central Maine Power	Barrier
Maine	Bangor Hydro Electric Company	Barrier
Maryland	BGE (Baltimore Gas and Electric)	Poor
Maryland	Potomac Electric Power Company	Poor
Massachusetts	NSTAR (Boston Edison)	Fair
Massachusetts	Massachusetts Electric Company	Fair
Michigan	Detroit Edison	Poor
Michigan	Consumers Energy Company	Poor
Minnesota	Xcel Energy (Northern States Power)	Fair
Mississippi	Entergy Mississippi (Mississippi Power and Light)	Barrier

Table A-19. (Continued).

State	Utility	Interconnection Policy Assessment
Mississippi	Mississippi Power Company	Barrier
Missouri	AmerenUE - Missouri (Union Electric)	Barrier
Montana	Northwestern Energy (Montana Power Company)	Poor
Nebraska	Omaha Public Power District	Barrier
Nevada	Nevada Power	Good
Nevada	Sierra Pacific Power Company	Good
New Hampshire	Public Service of New Hampshire	Poor
New Hampshire	Unitil Energy Systems	Poor
New Jersey	PSE&G (Public Service Electric and Gas Co.)	Good
New Jersey	Jersey Central Power and Light Co.	Good
New Jersey	Atlantic City Electrical Company	Good
New Mexico	PNM (Public Service Company of New Mexico)	Fair
New Mexico	Southwest Public Service Company	Fair
New York	Niagara Mohawk	Fair
New York	New York State Electric and Gas Corp	Fair
New York	Consolidated Edison	Fair
New York	Long Island Power Authority	Fair
North Carolina	Duke Power	Barrier
North Carolina	Progress Energy Carolinas Inc	Barrier
North Dakota	Northern States Power Co	Poor
Ohio	Ohio Power Company	Fair
Ohio	Ohio Edison	Fair
Ohio	Cincinnati Gas & Electric Company	Fair
Oklahoma	AEP (Public Service Company of Oklahoma)	Poor
Oklahoma	Oklahoma Gas and Electric Company	Poor
Oregon	PacifiCorp (Pacific Power)	Fair
Oregon	Portland General Electric Company	Fair
Pennsylvania	PPL Electric Utilities	Poor
Pennsylvania	PECO Energy Co	Poor
Pennsylvania	West Penn Power Co.	Poor
Rhode Island	Narragansett Electric	Poor
South Carolina	South Carolina Electric and Gas	Poor
South Carolina	Duke Energy Corporation	Poor
South Dakota	Xcel Energy (Northern States Power)	Barrier
Tennessee	Nashville Electric Service	Barrier
Tennessee	Knoxville Electric Board	Barrier

State	Utility	Interconnection Policy Assessment
Tennessee	City of Memphis	Barrier
Texas	TXU Electric	Fair
Texas	Reliant Energy Services	Fair
Texas	Entergy Gulf States Inc	Fair
Texas	Constellation New Energy Inc	Fair
Texas	City of San Antonio	Fair
Utah	PacifiCorp (Utah Power & Light)	Barrier
Vermont	Green Mountain Power	Fair
Vermont	Central Vermont Public Service Corporation	Fair
Virginia	Dominion (Virginia Electric and Power)	Poor
Virginia	Appalachian Power Co	Poor
Washington	Puget Sound Energy	Barrier
Washington	Snohomish County PUD No 1	Barrier
Washington	City of Seattle	Barrier
Washington, DC	PEPCO	Barrier
West Virginia	American Electric (Appalachian Power)	Poor
Wisconsin	We Energies (Wisconsin Electric)	Poor
Wisconsin	Wisconsin Public Service Corporation	Poor
Wyoming	PacifiCorp (Pacific Power)	Barrier

Rate Structures

NCI researched each utility's Web site to locate residential and commercial electric rates. We then confirmed with the Federal Energy Regulatory Commission (FERC) Form 1 Database about which standard and TOU rates are most representative of that utility. There are up to three rate structures for each utility's residential and commercial electric services: (1) standard; (2) TOU, weekday (if TOU is available); (3) TOU, weekend (if TOU is available). For each representative utility and assumed system size, we looked at TOU and standard rates to see which rate would yield a lower annual electric utility bill (with PV). We then used that rate structure for the analysis. Refer to the model for actual rate structures.

Demand Charges

NCI cataloged utility peak demand charges from utility Web sites and tariff sheets. We assumed that PV offsets only peak demand charges.

State and Local Incentives

NCI's PV Services Program provided a comprehensive list of local incentives for PV, broken down by state or utility. We divided incentives into three types: capacity-based (in $/kW), performance-based, and capacity-based (as a percentage of system cost). We found out when program funding was scheduled to run out and integrated that into the model. In cases where data could not be found, we implemented a switch to allow incentives to expire in 2009, 2012, or 2016. All the analysis performed in the study assumed the year to be 2009, to be conservative. In reality, if tax credits are extended, most state-level subsidies will be reduced or eliminated. Given that all cases analyzed, except the worst case, assume that federal tax credits are extended, we believe this is a good assumption.

For the California Solar Initiative, we implemented a feedback mechanism in the model that mimics the actual feedback mechanism being used in the initiative. In other words, when cumulative installations within a utility service area reach a certain level, the rebate amount is reduced. However, this model reduces the incentives on an annual basis only, rather than continuously.

Five-Year MACRS Depreciation

We amortized Modified Accelerated Cost-Recovery System (MACRS) benefits over the system life to account for the benefits of accelerated depreciation within the context of a modified simple payback in the commercial sector.

Net Metering Rules

NCI catalogued net metering rules for each state (or utility, where applicable) and accounted for the following: (1) Is net metering allowed? (2) If so, at what rate is electricity sold back to the grid? (3) Can customers get credit for electricity sold back in excess of their annual bill? (4) If so, at what rate is excess credit bought? Options for sell- back include retail, wholesale, and annual average rate. We collected data on these rates where necessary from EIA and internal NCI sources.

Table A-20. Net Metering Availability and Sell-Back Rules for Representative Utilities Analyzed

State	Utility	Net Metering Allowed?	Net Metering Sell Back Rates
Alabama	Alabama Power Co.	N	
Alaska	Chugach	N	
Arizona	Arizona Public Service	Y	Retail
Arizona	Salt River Project	Y	Retail
Arizona	Tucson Electric Power	Y	Retail
Arkansas	Entergy Arkansas	Y	Retail
California	Southern California Edison	Y	Retail
California	Sacramento Municipal Utility District	Y	Retail
California	Pacific Gas and Electric Company	Y	Retail
California	San Diego Gas & Electric Company	Y	Retail
California	Los Angeles Department of Water and Power	Y	Retail
Colorado	Public Service Company of Colorado	Y	Retail
Colorado	Colorado Springs	Y	Retail
Connecticut	Connecticut Light and Power	Y	Retail
Delaware	Conective (Delmarva Power)	Y	Retail
Florida	Florida Power & Light Co.	N	

State	Utility	Net Metering Allowed?	Net Metering Sell Back Rates
Florida	Progress Energy Florida Inc	N	
Florida	Tampa Electric Company	N	
Georgia	Georgia Power	Y	Retail
Hawaii	Hawaiian Electric Company (Oahu)	Y	Retail
Hawaii	Maui Electric Company	Y	Retail
Idaho	Idaho Power	Y	Retail
Illinois	Commonwealth Edison Co.	Y	Retail
Illinois	Illinois Power Company	N	
Indiana	PSI Energy Inc.	Y	Retail
Iowa	IES Utilities (mid america)	Y	Retail
Iowa	Interstate Power and Light	Y	Retail
Kansas	Kansas Gas & Electric Co	N	
Kansas	Westar Energy Inc	N	
Kentucky	Kentucky Utilities Co	Y	Retail
Kentucky	Louisville Gas & Electric Co	Y	Retail
Kentucky	Kenergy Corporation	Y	Retail
Louisiana	Entergy (Louisiana Power & Light)	Y	Retail
Maine	Central Maine Power	Y	Retail
Maine	Bangor Hydro Electric Company	Y	Retail
Maryland	BGE (Baltimore Gas and Electric)	Y	Retail
Maryland	Potomac Electric Power Company	Y	Retail
Massachusetts	NSTAR (Boston Edison)	Y	Retail
Massachusetts	Massachusetts Electric Company	Y	Retail
Michigan	Detroit Edison	Y	Retail
Michigan	Consumers Energy Company	Y	Retail
Minnesota	Xcel Energy (Northern States Power)	Y	Retail
Mississippi	Entergy Mississippi (Mississippi Power and Light)	N	
Mississippi	Mississippi Power Company	N	
Missouri	AmerenUE - Missouri (Union Electric)	Y	Wholesale
Montana	Northwestern Energy (Montana Power Company)	Y	Retail
Nebraska	Omaha Public Power District	N	
Nevada	Nevada Power	Y	Retail
Nevada	Sierra Pacific Power Company	Y	Retail
New Hampshire	Public Service of New Hampshire	Y	Retail
New Hampshire	Unitil Energy Systems	Y	Retail
New Jersey	PSE&G (Public Service Electric and Gas Co.)	Y	Retail

Table A-20. (Continued).

State	Utility	Net Metering Allowed?	Net Metering Sell Back Rates
New Jersey	Jersey Central Power and Light Co.	Y	Retail
New Jersey	Atlantic City Electrical Company	Y	Retail
New Mexico	PNM (Public Service Company of New Mexico)	Y	Retail
New Mexico	Southwest Public Service Company	Y	Retail
New York	Niagara Mohawk	Y	Retail
New York	New York State Electric and Gas Corp	Y	Retail
New York	Consolidated Edison	Y	Retail
New York	Long Island Power Authority	Y	Retail
North Carolina	Duke Power	Y	Retail
North Carolina	Progress Energy Carolinas Inc	Y	Retail
North Dakota	Northern States Power Co	Y	Wholesale
Ohio	Ohio Power Company	Y	Wholesale
Ohio	Ohio Edison	Y	Wholesale
Ohio	Cincinatti Gas & Electric Company	Y	Wholesale
Oklahoma	AEP (Public Service Company of Oklahoma)	Y	Retail
Oklahoma	Oklahoma Gas and Electric Company	Y	Retail
Oregon	PacifiCorp (Pacific Power)	Y	Retail
Oregon	Portland General Electric Company	Y	Retail
Pennsylvania	PPL Electric Utilities	Y	Retail
Pennsylvania	PECO Energy Co	Y	Retail
Pennsylvania	West Penn Power Co.	Y	Retail
Rhode Island	Narragansett Electric	Y	Retail
South Carolina	South Carolina Electric and Gas	N	
South Carolina	Duke Energy Corporation	N	
South Dakota	Xcel Energy (Northern States Power)	N	
Tennessee	Nashville Electric Service	N	
Tennessee	Knoxville Electric Board	N	
Tennessee	City of Memphis	N	
Texas	TXU Electric	Y	Retail
Texas	Reliant Energy Services	Y	Retail
Texas	Entergy Gulf States Inc	Y	Retail
Texas	Constellation New Energy Inc	Y	Retail
Texas	City of San Antonio	Y	Retail
Utah	PacifiCorp (Utah Power & Light)	Y	Retail
Vermont	Green Mountain Power	Y	Retail

State	Utility	Net Metering Allowed?	Net Metering Sell Back Rates
Vermont	Central Vermont Public Service Corporation	Y	Retail
Virginia	Dominion (Virginia Electric and Power)	Y	Retail
Virginia	Appalachian Power Co	Y	Retail
Washington	Puget Sound Energy	Y	Retail
Washington	Snohomish County PUD No 1	Y	Retail
Washington	City of Seattle	Y	Retail
Washington, DC	PEPCO	Y	Retail
West Virginia	American Electric (Appalachian Power)	Y	Retail
Wisconsin	We Energies (Wisconsin Electric)	Y	Retail
Wisconsin	Wisconsin Public Service Corporation	Y	Retail
Wyoming	PacifiCorp (Pacific Power)	Y	Retail

Table A-21. Net Metering Caps for Representative Utilities Analyzed

Utility	Do Net Metering Caps exist?	Cap Amount (% of utilities peak demand unless otherwise noted)
Alabama Power Co.	N	
Chugach	N	
Arizona Public Service	N	
Salt River Project	N	
Tucson Electric Power	N	
Entergy Arkansas	N	
Southern California Edison	Y	2.50%
Sacramento Municipal Utility District	Y	2.50%
Pacific Gas and Electric Company	Y	2.50%
San Diego Gas & Electric Company	Y	2.50%
Los Angeles Department of Water and Power	Y	2.50%
Public Service Company of Colorado	N	
Colorado Springs	N	
Connecticut Light and Power	N	
Conective (Delmarva Power)	N	
Florida Power & Light Co.	N	
Progress Energy Florida Inc	N	
Tampa Electric Company	N	
Georgia Power	Y	0.2%
Hawaiian Electric Company (Oahu)	Y	0.5%
Maui Electric Company	Y	0.5%

Table A-21. (Continued).

Utility	Do Net Metering Caps exist?	Cap Amount (% of utilities peak demand unless otherwise noted)
Idaho Power	Y	0.1% Of 2000 peak demand
Commonwealth Edison Co.	N	
Illinois Power Company	N	
PSI Energy Inc.	Y	0.10%
IES Utilities (mid america)	N	
Interstate Power and Light	N	
Kansas Gas & Electric Co	N	
Westar Energy Inc	N	
Kentucky Utilities Co	Y	0.10%
Louisville Gas & Electric Co	Y	0.10%
Kenergy Corporation	Y	0.10%
Entergy (Louisiana Power & Light)	N	
Central Maine Power	N	
Bangor Hydro Electric Company	N	
BGE (Baltimore Gas and Electric)	Y	Fixed # of MW's
Potomac Electric Power Company	Y	Fixed # of MW's
NSTAR (Boston Edison)	N	
Massachusetts Electric Company	N	
Detroit Edison	Y	0.1%
Consumers Energy Company	Y	0.1%
Xcel Energy (Northern States Power)	N	
Entergy Mississippi (Mississippi Power and Light)	N	
Mississippi Power Company	N	
AmerenUE - Missouri (Union Electric)	Y	5.0%
Northwestern Energy (Montana Power Company)	N	
Omaha Public Power District	N	
Nevada Power	Y	1.0%
Sierra Pacific Power Company	Y	1.0%
Public Service of New Hampshire	Y	0.1%
Unitil Energy Systems	Y	0.1%
PSE&G (Public Service Electric and Gas Co.)	N	
Jersey Central Power and Light Co.	N	
Atlantic City Electrical Company	N	
PNM (Public Service Company of New Mexico)	N	
Southwest Public Service Company	N	

Utility	Do Net Metering Caps exist?	Cap Amount (% of utilities peak demand unless otherwise noted)
Niagara Mohawk	Y	0.1%
New York State Electric and Gas Corp	Y	0.1%
Consolidated Edison	Y	0.1%
Long Island Power Authority	Y	0.1%
Duke Power	Y	0.2%
Progress Energy Carolinas Inc	Y	0.2%
Northern States Power Co	N	
Ohio Power Company	Y	1.0%
Ohio Edison	Y	1.0%
Cincinatti Gas & Electric Company	Y	1.0%
AEP (Public Service Company of Oklahoma)	N	
Oklahoma Gas and Electric Company	N	
PacifiCorp (Pacific Power)	Y	0.5%
Portland General Electric Company	Y	0.5%
PPL Electric Utilities	N	
PECO Energy Co	N	
West Penn Power Co.	N	
Narragansett Electric	Y	Fixed # of MW's
South Carolina Electric and Gas	N	
Duke Energy Corporation	N	
Xcel Energy (Northern States Power)	N	
Nashville Electric Service	N	
Knoxville Electric Board	N	
City of Memphis	N	
TXU Electric	N	
Reliant Energy Services	N	
Entergy Gulf States Inc	N	
Constellation New Energy Inc	N	
City of San Antonio	N	
PacifiCorp (Utah Power & Light)	Y	0.1% of 2001 peak demand
Green Mountain Power	Y	1.0%
Central Vermont Public Service Corporation	Y	1.0%
Dominion (Virginia Electric and Power)	Y	0.1%
Appalachian Power Co	Y	0.1%
Puget Sound Energy	N	0.25% of 1996 peak
Snohomish County PUD No 1	N	0.25% of 1996 peak
City of Seattle	N	0.25% of 1996 peak

Table A-21. (Continued).

Utility	Do Net Metering Caps exist?	Cap Amount (% of utilities peak demand unless otherwise noted)
PEPCO	N	
American Electric (Appalachian Power)	Y	0.1%
We Energies (Wisconsin Electric)	N	
Wisconsin Public Service Corporation	N	
PacifiCorp (Pacific Power)	N	

REC Assumptions

NCI cataloged current renewable energy credit (REC) prices in existing REC markets. For states with an RPS that have not established a REC market, we used a REC value of 15% below the alternative compliance payment. For those states, we assumed a REC market is partially developed in 2009 and fully developed in 2010. For states with separate solar alternative compliance payments, we assumed that if, in the previous year of analysis, the RPS solar set-aside target is met for the current year, the market value of a REC drops to 15% below the normal alternative compliance payment level for the current year (which is necessary only in the District of Columbia, Delaware, Maryland, New Jersey, and Pennsylvania). More refined methods cannot be used because the model has a temporal resolution of only one year.

Building Load Profiles

For residential buildings, NREL provided 8,760 building load profiles on a regional basis using weather for 2003 as an input. NCI and NREL identified 10 representative cities. We then assigned each utility a representative load profile based upon the utility's climate zone, as specified by Building America. The 15 cities were Phoenix, Sacramento, Los Angeles, Boulder, Tampa, Atlanta, Chicago, New York City, Houston, Seattle, Honolulu, Lexington, Dallas, Medford, and Helena.

For commercial buildings, NREL provided 8,760 building load profiles for all 98 utilities being analyzed, using weather data for 2003. Typical building load profiles were for office buildings, warehouses, or hospitals.

PV Output Profiles

For residential buildings, NREL provided 8,760 PV output profiles on a regional basis using 2003 weather as an input into PV Watts with a 30-degree tilt. NCI and NREL identified 15 representative cities. We then assigned each utility a representative PV system output profile.

For commercial buildings, NREL provided 8,760 PV output profiles for all 98 utilities being analyzed, using 2003 weather data as an input to PV Watts with a 0-degree tilt.

O&M and Inverter Costs

DOE provided NCI with aggregated, combined O&M and inverter replacement costs from applicants and awardees of the Solar America Initiative.

Table A-22. O&M and Inverter Replacement Costs

Market Segment	O&M Costs and Inverter Replacement Costs ($kW/yr)		
	2007	2010	2015
Residential	$57.98	$39.45	$35.00
Commercial	$51.28	$38.07	$27.33

System Size

NCI started with default system sizing of 5 kW in the residential sector and 250 kW in the commercial sector. We then reduced system size based on net-metering rules, interconnection standards and local incentive amounts to maximize the value of the incentive (i.e., if a utility offers rebates only for the first 100 kW, a 1 00-kW system size was used).

Calculation of Annual Electric Bill Savings

Using 8,760 building load profiles provided by NREL and actual utility rate structures (accounting for seasonal variation, TOU rates, and so on), first we calculated a customer's annual electric bill. Next, we calculated annual electric bill savings by combining 8,760 PV output profiles, actual utility rate structures, and the local net-metering laws (i.e., whether net metering is allowed, the rate at which power is sold back to the grid, and whether a customer can sell back power in excess of their annual electric bill).

Information on Calculated TOU Rates

Not all state utility rates used in the analysis conform nicely to average TOU structures. Where applicable, extreme outliers were ignored in the calculation. For example, PSI Energy, Inc., was ignored in the analysis of the Reliability*First* Corporation (RFC) region because its existing TOU rate is available only to those customers with its low-load factor service, a very specific rate. Within the Northeast Power Coordinating Council (NPCC) region, Central Maine Power is the only utility with a shoulder period and rate; thus, a weighted average of the peak and shoulder rates and times was taken to create a new, representative peak rate and length of time.

As expected, TOU structures tended to vary within each region. For example, Florida utilities all establish a morning peak and an evening peak period with nonpeak rates throughout the middle of the day. The average changes in peak-hour rates and non-peak-hour rates between the the winter and summer seasons vary the most between the Northeast (NE) and Pacific states; the NE shows almost no change between seasons, and the Southwest and West show as much as a 147% increase in commercial peak rates between the two seasons. The utility structures within the RFC region vary the most, potentially as a result of the recent merger of the East Central Area Reliability Coordination Agreement (ECAR), the Mid-Atlantic Area Council (MAAC), and the Mid-America Interconnected Network (MAIN) regional reliability councils.

Impact of Carbon Pricing

To examine the impacts of potential national carbon legislation, we modeled the price of carbon as a surcharge on retail electric rates. To assess the impact on electric rates, we used carbon intensity data from EIA's Annual Energy Outlook, by EMMR, and developed $/kWh impacts for $/ton pricing. See below for the values calculated.

Table A-23. Impact of Carbon Cap

Utility IDs	Utility Names	Impact of Carbon Cap [$/kWh per $/ton]
1	Alabama Power Co.	0.00058
2	Chugach	0.00016
3	Arizona Public Service	0.00064
4	Salt River Project	0.00064
5	Tucson Electric Power	0.00064
6	Entergy Arkansas	0.00058
7	Southern California Edison	0.00031
8	Sacramento Municipal Utility District	0.00031
9	Pacific Gas and Electric Company	0.00031
10	San Diego Gas & Electric Company	0.00031
11	Los Angeles Department of Water and Power	0.00031
12	Public Service Company of Colorado	0.00064
13	Colorado Springs	0.00064
14	Connecticut Light and Power	0.00039
15	Conective (Delmarva Power)	0.00051
16	Florida Power & Light Co.	0.00057
17	Progress Energy Florida Inc	0.00057
18	Tampa Electric Company	0.00057
19	Georgia Power	0.00058
20	Hawaiian Electric Company (Oahu)	0.00016
21	Maui Electric Company	0.00016
22	Idaho Power	0.00037
23	Commonwealth Edison Co.	0.00060
24	Illinois Power Company	0.00060
25	PSI Energy Inc.	0.00083
26	IES Utilities (mid america)	0.00060
27	Interstate Power and Light	0.00060
28	Kansas Gas & Electric Co	0.00084
29	Westar Energy Inc	0.00084
30	Kentucky Utilities Co	0.00083
31	Louisville Gas & Electric Co	0.00083
32	Kenergy Corporation	0.00083
33	Entergy (Louisiana Power & Light)	0.00058

Utility IDs	Utility Names	Impact of Carbon Cap [$/kWh per $/ton]
34	Central Maine Power	0.00039
35	Bangor Hydro Electric Company	0.00039
36	BGE (Baltimore Gas and Electric)	0.00051
37	Potomac Electric Power Company	0.00051
38	NSTAR (Boston Edison)	0.00039
39	Massachusetts Electric Company	0.00039
40	Detroit Edison	0.00083
41	Consumers Energy Company	0.00083
42	Xcel Energy (Northern States Power)	0.00077
43	Entergy Mississippi (Mississippi Power and Light)	0.00058
44	Mississippi Power Company	0.00058
45	AmerenUE - Missouri (Union Electric)	0.00060
46	Northwestern Energy (Montana Power Company)	0.00037
47	Omaha Public Power District	0.00077
48	Nevada Power	0.00037
49	Sierra Pacific Power Company	0.00037
50	Public Service of New Hampshire	0.00039
51	Unitil Energy Systems	0.00039
52	PSE&G (Public Service Electric and Gas Co.)	0.00051
53	Jersey Central Power and Light Co.	0.00051
54	Atlantic City Electrical Company	0.00051
55	PNM (Public Service Company of New Mexico)	0.00064
56	Southwest Public Service Company	0.00064
57	Niagara Mohawk	0.00033
58	New York State Electric and Gas Corp	0.00033
59	Consolidated Edison	0.00033
60	Long Island Power Authority	0.00033
61	Duke Power	0.00058
62	Progress Energy Carolinas Inc	0.00058
63	Northern States Power Co	0.00077
64	Ohio Power Company	0.00083
65	Ohio Edison	0.00083
66	Cincinnati Gas & Electric Company	0.00083
67	AEP (Public Service Company of Oklahoma)	0.00084
68	Oklahoma Gas and Electric Company	0.00084
69	PacifiCorp (Pacific Power)	0.00037
70	Portland General Electric Company	0.00037

Table A-23. (Continued).

Utility IDs	Utility Names	Impact of Carbon Cap [$/kWh per $/ton]
71	PPL Electric Utilities	0.00051
72	PECO Energy Co	0.00051
73	West Penn Power Co.	0.00051
74	Narragansett Electric	0.00039
75	South Carolina Electric and Gas	0.00058
76	Duke Energy Corporation	0.00058
77	Xcel Energy (Northern States Power)	0.00077
78	Nashville Electric Service	0.00058
79	Knoxville Electric Board	0.00058
80	City of Memphis	0.00058
81	TXU Electric	0.00057
82	Reliant Energy Services	0.00057
83	Entergy Gulf States Inc	0.00057
84	Constellation New Energy Inc	0.00057
85	City of San Antonio	0.00057
86	PacifiCorp (Utah Power & Light)	0.00037
87	Green Mountain Power	0.00039
88	Central Vermont Public Service Corporation	0.00039
89	Dominion (Virginia Electric and Power)	0.00058
90	Appalachian Power Co	0.00058
91	Puget Sound Energy	0.00037
92	Snohomish County PUD No 1	0.00037
93	City of Seattle	0.00037
94	PEPCO	0.00051
95	American Electric (Appalachian Power)	0.00083
96	We Energies (Wisconsin Electric)	0.00060
97	Wisconsin Public Service Corporation	0.00060
98	PacifiCorp (Pacific Power)	0.00037

Electricity Escalation Rates

We used two rate escalation scenarios, all in real terms. EIA's Annual Energy Outlook provided the first, but it shows rates staying constant or dropping in all markets. As a result, NCI conducted an analysis looking at projections of supply, capacity, and policy changes that will impact the annual wholesale price. NCI then assumed that changes in wholesale prices will be 100% translated to the retail market (the model allows the user to alter this function). This is a strong assumption, but looking at the dynamics between wholesale and retail markets is outside the scope of the project. The resulting annual percent changes in prices are shown in the tables that follow.

Table A-24. Annual Year Over Year Changes in Electricity Prices as Projected by EIA for the Residential Market

State	2008	2009	2010	2011	2012	2013	2014	2015
Alabama	-0.68%	0.05%	-0.02%	-1.26%	-2.02%	-1.17%	-0.61%	-0.39%
Alaska	-0.95%	-1.29%	-1.53%	-2.34%	-1.93%	-1.04%	-0.46%	0.35%
Arizona	0.08%	-1.83%	-1.85%	-2.29%	-0.24%	1.45%	-0.10%	-0.01%
Arkansas	-1.31%	-2.37%	-2.19%	-2.68%	1.18%	-0.10%	-0.20%	-0.52%
California	-0.95%	-1.29%	-1.53%	-2.34%	-1.93%	-1.04%	-0.46%	0.35%
Colorado	0.08%	-1.83%	-1.85%	-2.29%	-0.24%	1.45%	-0.10%	-0.01%
Connecticut	-1.98%	-0.12%	-0.06%	-1.29%	-1.01%	1.28%	-1.44%	1.61%
Delaware	0.11%	-0.88%	-0.48%	-1.80%	-2.03%	-1.20%	-0.78%	-0.34%
Florida	0.11%	-0.88%	-0.48%	-1.80%	-2.03%	-1.20%	-0.78%	-0.34%
Georgia	0.11%	-0.88%	-0.48%	-1.80%	-2.03%	-1.20%	-0.78%	-0.34%
Hawaii	-0.95%	-1.29%	-1.53%	-2.34%	-1.93%	-1.04%	-0.46%	0.35%
Idaho	0.08%	-1.83%	-1.85%	-2.29%	-0.24%	1.45%	-0.10%	-0.01%
Illinois	-0.89%	0.42%	-0.12%	0.12%	-1.18%	-0.12%	0.02%	0.03%
Indiana	-0.89%	0.42%	-0.12%	0.12%	-1.18%	-0.12%	0.02%	0.03%
Iowa	0.82%	1.45%	1.43%	-0.10%	-1.52%	-0.64%	-0.90%	-1.12%
Kansas	0.82%	1.45%	1.43%	-0.10%	-1.52%	-0.64%	-0.90%	-1.12%
Kentucky	-0.68%	0.05%	-0.02%	-1.26%	-2.02%	-1.17%	-0.61%	-0.39%
Louisiana	-1.31%	-2.37%	-2.19%	-2.68%	1.18%	-0.10%	-0.20%	-0.52%
Maine	-1.98%	-0.12%	-0.06%	-1.29%	-1.01%	1.28%	-1.44%	1.61%
Maryland	0.11%	-0.88%	-0.48%	-1.80%	-2.03%	-1.20%	-0.78%	-0.34%
Massachusetts	-1.98%	-0.12%	-0.06%	-1.29%	-1.01%	1.28%	-1.44%	1.61%
Michigan	-0.89%	0.42%	-0.12%	0.12%	-1.18%	-0.12%	0.02%	0.03%
Minnesota	0.82%	1.45%	1.43%	-0.10%	-1.52%	-0.64%	-0.90%	-1.12%
Mississippi	-0.68%	0.05%	-0.02%	-1.26%	-2.02%	-1.17%	-0.61%	-0.39%
Missouri	0.82%	1.45%	1.43%	-0.10%	-1.52%	-0.64%	-0.90%	-1.12%
Montana	0.08%	-1.83%	-1.85%	-2.29%	-0.24%	1.45%	-0.10%	-0.01%
Nebraska	0.82%	1.45%	1.43%	-0.10%	-1.52%	-0.64%	-0.90%	-1.12%
Nevada	0.08%	-1.83%	-1.85%	-2.29%	-0.24%	1.45%	-0.10%	-0.01%
New Hampshire	-1.98%	-0.12%	-0.06%	-1.29%	-1.01%	1.28%	-1.44%	1.61%
New Jersey	0.14%	-0.77%	-1.99%	-0.84%	1.67%	0.34%	-0.39%	0.09%
New Mexico	0.08%	-1.83%	-1.85%	-2.29%	-0.24%	1.45%	-0.10%	-0.01%
New York	0.14%	-0.77%	-1.99%	-0.84%	1.67%	0.34%	-0.39%	0.09%
North Carolina	0.11%	-0.88%	-0.48%	-1.80%	-2.03%	-1.20%	-0.78%	-0.34%
North Dakota	0.82%	1.45%	1.43%	-0.10%	-1.52%	-0.64%	-0.90%	-1.12%
Ohio	-0.89%	0.42%	-0.12%	0.12%	-1.18%	-0.12%	0.02%	0.03%
Oklahoma	-1.31%	-2.37%	-2.19%	-2.68%	1.18%	-0.10%	-0.20%	-0.52%
Oregon	-0.95%	-1.29%	-1.53%	-2.34%	-1.93%	-1.04%	-0.46%	0.35%
Pennsylvania	0.14%	-0.77%	-1.99%	-0.84%	1.67%	0.34%	-0.39%	0.09%
Rhode Island	-1.98%	-0.12%	-0.06%	-1.29%	-1.01%	1.28%	-1.44%	1.61%

State								
South Carolina	0.82%	1.45%	1.43%	-0.10%	-1.52%	-0.64%	-0.90%	-1.12%
South Dakota	0.11%	-0.88%	-0.48%	-1.80%	-2.03%	-1.20%	-0.78%	-0.34%
Tennessee	-0.68%	0.05%	-0.02%	-1.26%	-2.02%	-1.17%	-0.61%	-0.39%
Texas	-1.31%	-2.37%	-2.19%	-2.68%	1.18%	-0.10%	-0.20%	-0.52%
Utah	0.08%	-1.83%	-1.85%	-2.29%	-0.24%	1.45%	-0.10%	-0.01%
Vermont	-1.98%	-0.12%	-0.06%	-1.29%	-1.01%	1.28%	-1.44%	1.61%
Virginia	0.11%	-0.88%	-0.48%	-1.80%	-2.03%	-1.20%	-0.78%	-0.34%
Washington	-0.95%	-1.29%	-1.53%	-2.34%	-1.93%	-1.04%	-0.46%	0.35%
Washington, DC	0.11%	-0.88%	-0.48%	-1.80%	-2.03%	-1.20%	-0.78%	-0.34%
West Virginia	0.11%	-0.88%	-0.48%	-1.80%	-2.03%	-1.20%	-0.78%	-0.34%
Wisconsin	-0.89%	0.42%	-0.12%	0.12%	-1.18%	-0.12%	0.02%	0.03%
Wyoming	0.08%	-1.83%	-1.85%	-2.29%	-0.24%	1.45%	-0.10%	-0.01%

Table A-25. Annual Year-Over-Year Changes in Electricity Prices as Projected by EIA for the Commercial Market

State	2008	2009	2010	2011	2012	2013	2014	2015
Alabama	-0.02%	0.33%	0.25%	-1.51%	-2.11%	-1.30%	-0.49%	-0.20%
Alaska	-0.80%	-1.90%	-2.27%	-2.90%	-2.34%	-1.47%	-0.69%	0.15%
Arizona	0.36%	-2.20%	-2.56%	-3.50%	-0.62%	1.89%	0.00%	0.10%
Arkansas	0.07%	-2.02%	-2.20%	-3.27%	0.75%	-0.32%	-0.15%	-0.27%
California	-0.80%	-1.90%	-2.27%	-2.90%	-2.34%	-1.47%	-0.69%	0.15%
Colorado	0.36%	-2.20%	-2.56%	-3.50%	-0.62%	1.89%	0.00%	0.10%
Connecticut	-2.96%	-2.81%	-2.36%	-4.29%	-1.95%	0.91%	-1.19%	2.71%
Delaware	0.58%	-0.88%	-0.63%	-2.14%	-1.80%	-1.12%	-0.63%	-0.12%
Florida	0.58%	-0.88%	-0.63%	-2.14%	-1.80%	-1.12%	-0.63%	-0.12%
Georgia	0.58%	-0.88%	-0.63%	-2.14%	-1.80%	-1.12%	-0.63%	-0.12%
Hawaii	-0.80%	-1.90%	-2.27%	-2.90%	-2.34%	-1.47%	-0.69%	0.15%
Idaho	0.36%	-2.20%	-2.56%	-3.50%	-0.62%	1.89%	0.00%	0.10%
Illinois	-0.04%	0.35%	-0.50%	-1.35%	-2.48%	-0.95%	-0.34%	0.07%
Indiana	-0.04%	0.35%	-0.50%	-1.35%	-2.48%	-0.95%	-0.34%	0.07%
Iowa	1.61%	1.72%	1.53%	-0.32%	-1.94%	-0.84%	-0.92%	-1.04%
Kansas	1.61%	1.72%	1.53%	-0.32%	-1.94%	-0.84%	-0.92%	-1.04%
Kentucky	-0.02%	0.33%	0.25%	-1.51%	-2.11%	-1.30%	-0.49%	-0.20%
Louisiana	0.07%	-2.02%	-2.20%	-3.27%	0.75%	-0.32%	-0.15%	-0.27%
Maine	-2.96%	-2.81%	-2.36%	-4.29%	-1.95%	0.91%	-1.19%	2.71%
Maryland	0.58%	-0.88%	-0.63%	-2.14%	-1.80%	-1.12%	-0.63%	-0.12%
Massachusetts	-2.96%	-2.81%	-2.36%	-4.29%	-1.95%	0.91%	-1.19%	2.71%
Michigan	-0.04%	0.35%	-0.50%	-1.35%	-2.48%	-0.95%	-0.34%	0.07%
Minnesota	1.61%	1.72%	1.53%	-0.32%	-1.94%	-0.84%	-0.92%	-1.04%
Mississippi	-0.02%	0.33%	0.25%	-1.51%	-2.11%	-1.30%	-0.49%	-0.20%
Missouri	1.61%	1.72%	1.53%	-0.32%	-1.94%	-0.84%	-0.92%	-1.04%

Montana	0.36%	-2.20%	-2.56%	-3.50%	-0.62%	1.89%	0.00%	0.10%
Nebraska	1.61%	1.72%	1.53%	-0.32%	-1.94%	-0.84%	-0.92%	-1.04%
Nevada	0.36%	-2.20%	-2.56%	-3.50%	-0.62%	1.89%	0.00%	0.10%
New Hampshire	-2.96%	-2.81%	-2.36%	-4.29%	-1.95%	0.91%	-1.19%	2.71%
New Jersey	-1.17%	-3.63%	-4.75%	-4.13%	1.83%	-0.22%	-0.67%	0.47%
New Mexico	0.36%	-2.20%	-2.56%	-3.50%	-0.62%	1.89%	0.00%	0.10%
New York	-1.17%	-3.63%	-4.75%	-4.13%	1.83%	-0.22%	-0.67%	0.47%
North Carolina	0.58%	-0.88%	-0.63%	-2.14%	-1.80%	-1.12%	-0.63%	-0.12%
North Dakota	1.61%	1.72%	1.53%	-0.32%	-1.94%	-0.84%	-0.92%	-1.04%
Ohio	-0.04%	0.35%	-0.50%	-1.35%	-2.48%	-0.95%	-0.34%	0.07%
Oklahoma	0.07%	-2.02%	-2.20%	-3.27%	0.75%	-0.32%	-0.15%	-0.27%
Oregon	-0.80%	-1.90%	-2.27%	-2.90%	-2.34%	-1.47%	-0.69%	0.15%
Pennsylvania	-1.17%	-3.63%	-4.75%	-4.13%	1.83%	-0.22%	-0.67%	0.47%
Rhode Island	-2.96%	-2.81%	-2.36%	-4.29%	-1.95%	0.91%	-1.19%	2.71%
South Carolina	1.61%	1.72%	1.53%	-0.32%	-1.94%	-0.84%	-0.92%	-1.04%
South Dakota	0.58%	-0.88%	-0.63%	-2.14%	-1.80%	-1.12%	-0.63%	-0.12%
Tennessee	-0.02%	0.33%	0.25%	-1.51%	-2.11%	-1.30%	-0.49%	-0.20%
Texas	0.07%	-2.02%	-2.20%	-3.27%	0.75%	-0.32%	-0.15%	-0.27%
Utah	0.36%	-2.20%	-2.56%	-3.50%	-0.62%	1.89%	0.00%	0.10%
Vermont	-2.96%	-2.81%	-2.36%	-4.29%	-1.95%	0.91%	-1.19%	2.71%
Virginia	0.58%	-0.88%	-0.63%	-2.14%	-1.80%	-1.12%	-0.63%	-0.12%
Washington	-0.80%	-1.90%	-2.27%	-2.90%	-2.34%	-1.47%	-0.69%	0.15%
Washington, DC	0.58%	-0.88%	-0.63%	-2.14%	-1.80%	-1.12%	-0.63%	-0.12%
West Virginia	0.58%	-0.88%	-0.63%	-2.14%	-1.80%	-1.12%	-0.63%	-0.12%
Wisconsin	-0.04%	0.35%	-0.50%	-1.35%	-2.48%	-0.95%	-0.34%	0.07%
Wyoming	0.36%	-2.20%	-2.56%	-3.50%	-0.62%	1.89%	0.00%	0.10%

Table A-26. Annual Year-Over-Year Changes in Electricity Prices as Projected by NCI

State	2008	2009	2010	2011	2012	2013	2014	2015
Alabama	13.52%	2.76%	-5.60%	5.40%	7.72%	-0.01%	12.62%	3.13%
Alaska	0.29%	1.24%	1.47%	6.00%	-4.60%	-0.06%	-0.06%	-0.06%
Arizona	15.82%	-0.81%	-11.04%	10.69%	7.24%	0.59%	15.48%	0.78%
Arkansas	-9.78%	-1.38%	-2.97%	-0.75%	0.52%	-1.03%	14.80%	-8.39%
California	12.04%	-4.70%	-13.04%	9.91%	7.64%	-0.43%	14.75%	1.90%
Colorado	11.25%	-6.09%	-14.80%	9.68%	6.84%	-1.90%	13.77%	1.93%
Connecticut	8.85%	-1.09%	-8.47%	5.95%	7.73%	0.67%	13.70%	-4.63%
Delaware	11.99%	-3.40%	-7.53%	6.48%	6.56%	5.80%	14.99%	-3.38%
Florida	11.73%	1.73%	-11.50%	6.79%	4.32%	-1.18%	8.53%	-0.02%
Georgia	-9.03%	-0.92%	-4.26%	0.82%	0.68%	0.20%	14.29%	-7.25%
Hawaii	-0.32%	5.07%	-0.06%	-0.06%	-0.06%	-0.06%	-0.06%	-0.06%
Idaho	12.08%	-5.85%	-12.46%	5.00%	5.01%	-0.59%	13.71%	2.41%

Table A-26 (Continued)

Illinois	-1.78%	6.41%	-8.77%	-6.77%	-1.18%	2.17%	12.53%	9.08%
Indiana	-1.36%	6.51%	-7.83%	-7.13%	-1.18%	3.15%	11.95%	8.23%
Iowa	-11.84%	2.76%	6.04%	-1.85%	-1.44%	-1.39%	12.66%	-8.97%
Kansas	-11.16%	-2.55%	-2.04%	-1.02%	0.94%	-0.71%	17.98%	-7.99%
Kentucky	13.52%	2.76%	-5.60%	5.40%	7.72%	-0.01%	12.62%	3.13%
Louisiana	-9.78%	-1.38%	-2.97%	-0.75%	0.52%	-1.03%	14.80%	-8.39%
Maine	8.38%	-2.95%	-10.57%	5.68%	7.17%	1.11%	13.29%	-3.26%
Maryland	11.99%	-3.40%	-7.53%	6.48%	6.56%	5.80%	14.99%	-3.38%
Massachusetts	9.03%	-3.71%	-5.91%	5.89%	7.94%	0.49%	13.85%	-5.00%
Michigan	-9.73%	-0.06%	-2.73%	2.14%	0.44%	4.84%	13.77%	-6.53%
Minnesota	-11.84%	2.76%	6.04%	-1.85%	-1.44%	-1.39%	12.66%	-8.97%
Mississippi	13.52%	2.76%	-5.60%	5.40%	7.72%	-0.01%	12.62%	3.13%
Missouri	-11.84%	2.76%	2.01%	-1.92%	-1.50%	-1.44%	13.19%	-9.30%
Montana	11.35%	-5.48%	-12.17%	3.99%	5.32%	-0.03%	12.58%	1.90%
Nebraska	-11.84%	2.76%	6.04%	-1.85%	-1.44%	-1.39%	12.66%	-8.97%
Nevada	15.82%	-0.81%	-11.04%	10.69%	7.24%	0.59%	15.48%	0.78%
New Hampshire	10.81%	-4.02%	-9.85%	5.67%	7.16%	0.51%	15.68%	-6.59%
New Jersey	11.99%	-3.40%	-7.53%	10.17%	6.34%	2.47%	14.99%	-3.38%
New Mexico	15.82%	-0.81%	-11.04%	10.69%	7.24%	0.59%	15.48%	0.78%
New York	9.57%	-9.02%	-12.07%	4.35%	7.21%	0.51%	14.12%	-4.78%
North Carolina	-9.90%	0.22%	-3.46%	1.33%	0.93%	0.79%	14.20%	-6.54%
North Dakota	-11.84%	2.76%	6.04%	-1.85%	-1.44%	-1.39%	12.66%	-8.97%
Ohio	-1.36%	6.51%	-7.83%	-7.13%	-1.18%	3.15%	11.95%	8.23%
Oklahoma	-11.16%	-2.55%	-2.04%	-1.02%	0.94%	-0.71%	17.98%	-7.99%
Oregon	11.35%	-5.48%	-12.17%	5.42%	5.51%	-0.03%	12.98%	1.96%
Pennsylvania	12.55%	2.41%	-4.69%	6.27%	7.25%	4.90%	13.63%	2.89%
Rhode Island	7.87%	-3.17%	-8.44%	5.50%	8.06%	1.14%	13.67%	-4.49%
South Carolina	-9.90%	0.22%	-3.46%	1.33%	0.93%	0.79%	14.20%	-6.54%
South Dakota	-11.84%	2.76%	6.04%	-1.85%	-1.44%	-1.39%	12.66%	-8.97%
Tennessee	13.52%	2.76%	-5.60%	5.40%	7.72%	-0.01%	12.62%	3.13%
Texas	-3.56%	-1.87%	-12.18%	-10.19%	-8.05%	-7.65%	11.91%	9.54%
Utah	12.93%	-4.06%	-13.01%	10.20%	7.08%	-0.16%	14.33%	1.57%
Vermont	9.80%	-2.65%	-9.85%	3.30%	8.53%	0.51%	14.68%	-4.95%
Virginia	11.99%	-3.40%	-7.53%	6.48%	6.56%	2.55%	15.47%	-3.47%
Washington	11.35%	-5.48%	-12.17%	5.42%	5.51%	-0.03%	12.98%	1.96%
Washington, DC	11.99%	-3.40%	-7.53%	6.48%	6.56%	5.80%	14.99%	-3.38%
West Virginia	12.55%	2.41%	-4.69%	6.27%	7.25%	4.90%	13.63%	2.89%
Wisconsin	-11.09%	0.22%	1.52%	-0.20%	0.39%	0.12%	13.22%	-8.58%
Wyoming	11.35%	-5.48%	-12.17%	3.99%	5.32%	-0.03%	12.58%	1.90%

In: Renewable Energy Grid Integration
Editor: Marco H. Balderas

ISBN: 978-1-60741-324-0
© 2009 Nova Science Publishers, Inc.

Chapter 2

PRODUCTION COST MODELING FOR HIGH LEVELS OF PHOTOVOLTAICS PENETRATION[*]

P. Denholm, R. Margolis and J. Milford

PREFACE

Now is the time to plan for the integration of significant quantities of distributed renewable energy into the electricity grid. Concerns about climate change, the adoption of state-level renewable portfolio standards and incentives, and accelerated cost reductions are driving steep growth in U.S. renewable energy technologies. The number of distributed solar photovoltaic (PV) installations, in particular, is growing rapidly. As distributed PV and other renewable energy technologies mature, they can provide a significant share of our nation's electricity demand. However, as their market share grows, concerns about potential impacts on the stability and operation of the electricity grid may create barriers to their future expansion.

To facilitate more extensive adoption of renewable distributed electric generation, the U.S. Department of Energy launched the Renewable Systems Interconnection (RSI) study during the spring of 2007. This study addresses the technical and analytical challenges that must be addressed to enable high penetration levels of distributed renewable energy technologies. Because integration-related issues at the distribution system are likely to emerge first for PV technology, the RSI study focuses on this area. A key goal of the RSI study is to identify the research and development needed to build the foundation for a high-penetration renewable energy future while enhancing the operation of the electricity grid.

The RSI study consists of 15 reports that address a variety of issues related to distributed systems technology development; advanced distribution systems integration; system-level tests and demonstrations; technical and market analysis; resource assessment; and codes, standards, and regulatory implementation. The RSI reports are:

- Renewable Systems Interconnection: Executive Summary

---

[*] Excerpted from Technical Report NREL/TP-581 -42305, dated February 2008.

- Distributed Photovoltaic Systems Design and Technology Requirements
- Advanced Grid Planning and Operation
- Utility Models, Analysis, and Simulation Tools
- Cyber Security Analysis
- Power System Planning: Emerging Practices Suitable for Evaluating the Impact of High-Penetration Photovoltaics
- Distribution System Voltage Performance Analysis for High-Penetration Photovoltaics
- Enhanced Reliability of Photovoltaic Systems with Energy Storage and Controls
- Transmission System Performance Analysis for High-Penetration Photovoltaics
- Solar Resource Assessment
- Test and Demonstration Program Definition
- Photovoltaics Value Analysis
- Photovoltaics Business Models
- Production Cost Modeling for High Levels of Photovoltaic Penetration • Rooftop Photovoltaics Market Penetration Scenarios.

Addressing grid-integration issues is a necessary prerequisite for the long-term viability of the distributed renewable energy industry, in general, and the distributed PV industry, in particular. The RSI study is one step on this path. The Department of Energy is also working with stakeholders to develop a research and development plan aimed at making this vision a reality.

EXECUTIVE SUMMARY

Solar PV is being deployed in part to reduce dependence on fossil fuels for electricity use and associated emissions of greenhouse gases and criteria pollutants such as nitrous oxides (NO_x) and sulfur dioxide (SO2). Given the time-varying output of photovoltaic (PV) equipment, and the diverse set of electric generators in the power plant fleet, there is considerable uncertainty as to the actual benefits of PV in various regions.

This chapter uses a production cost modeling approach to evaluate the large scale interaction of solar electricity technologies with the existing and possible future grid, with a focus on displaced generation capacity, fuel saved, and emissions avoided by deploying varying levels of solar electric generation. This study established a PV penetration scenario in several regions in the western U.S. grid (the Western Electricity Coordinating Council – WECC) and simulates the response of the power plant fleet. While focusing on avoided fuels and emissions that result from PV deployment, this analysis also identifies areas of future research to increase understanding of benefits and impacts of large-scale PV deployment.

The simulations evaluated a series of PV penetrations in which 1% to 10% of the entire western interconnect's annual electrical energy is derived from PV. The PV is distributed based on an assumed market penetration scenario with higher penetration in the Southwest and California and lower penetration in the Northeastern part of the region.

Figure E-1 illustrates the simulated impact of the deployment of PV during a single day in California under five penetration scenarios. On this day, the deployment of PV reduces the generation primarily from natural gas-fired power plants (labeled CC for combined- cycle and CT for combustion turbine).

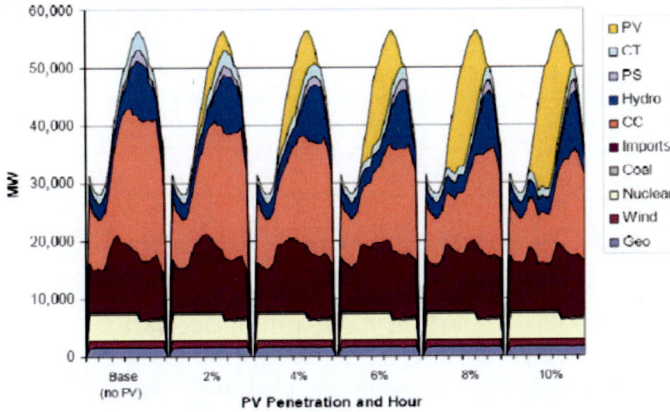

Figure E-1. Simulated Dispatch in California for a Summer Day in 2007 with Various PV Energy Penetration Scenarios.

Over the entire WECC region, PV displaces natural gas at low penetration, and begins to displace coal at higher penetration. Figure E-2 illustrates the average avoided fuel for each kWh of PV generation in the assumed scenario.

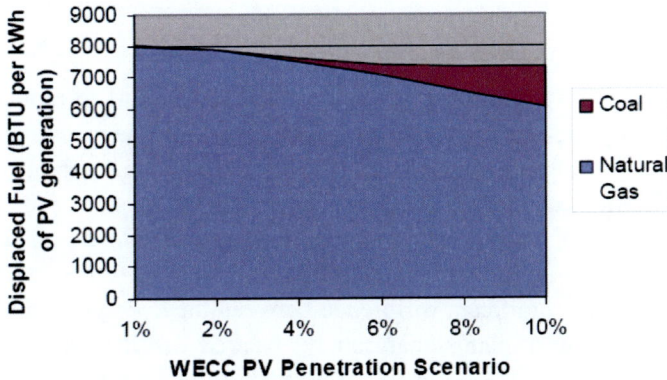

Figure E-2. Average Fuel Displacement Rate from PV Deployed in WECC.

The avoided emissions rate from PV depends on the fuel mix, and the changing generator efficiency as a function of load. Figure E-3 illustrates the average and marginal avoided carbon dioxide (CO_2) emissions rate for the assumed deployment scenario. (The average rate represents the emissions displacement rate for ALL PV generation at a specific penetration, while marginal rate represents the emissions displacement rate for the incremental unit of additional PV at a specific penetration level).

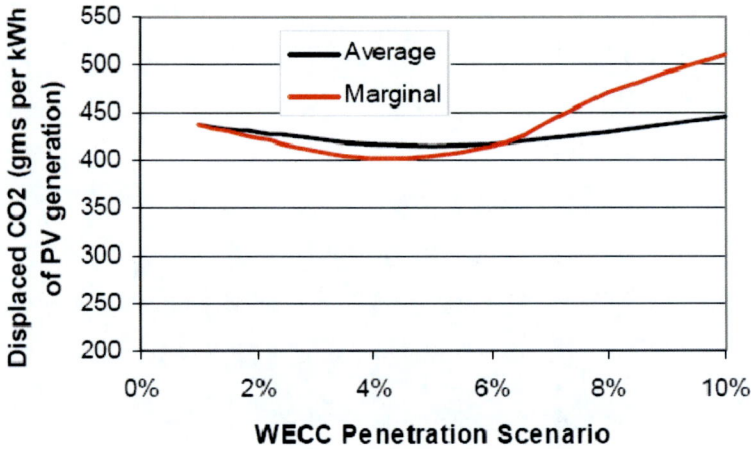

Figure E-3. Average and Marginal CO_2 Emissions Displacement from PV Deployed in WECC.

In addition to providing estimates of avoided fuels and emissions, this chapter also considers other analysis needed to evaluate grid-level impacts and benefits of distributed PV. Among these needs are evaluation of the integration costs of PV considering the effects of solar resource forecasting, the ability of generators to follow variations in PV output, decreased T&D losses, and capacity benefits.

1.0. INTRODUCTION

Solar photovoltaic (PV) technology is being deployed in part to reduce dependence on fossil fuels for electricity use and associated emissions of greenhouse gases and criteria pollutants such as nitrous oxides (NO_x) and sulfur dioxide (SO2). Given the time-varying output of PV, and the diverse set of electric generators in the power plant fleet, there is considerable uncertainty as to the actual benefits of PV in various regions. Simple grid- average emissions and fuel use provide unsatisfactory estimates of actual benefits given the peak-coincidence aspects of PV, along with the potentially significant difference between the average grid and the generators "on the margin." The power plants that can be backed off in response to the mid-day generation of PV electricity may be quite different from those providing constant baseload power.

This chapter uses a production cost modeling approach to evaluate the large scale interaction of solar electricity technologies with the existing and possible future grid, with a focus on displaced generation capacity, fuel saved, and emissions avoided by deploying varying levels of solar electric generation. This study established a PV penetration scenario in several regions in the United States and simulates the response of the power plant fleet. While focusing on avoided fuels and emissions that result from PV deployment, this analysis also identifies areas of future research to increase understanding of benefits and impacts of large-scale PV deployment.

2.0. CURRENT STATUS OF EXISTING RESEARCH

There are a number of approaches used to estimate the displaced fuels and emissions associated with the deployment of renewable energy technologies. The most basic approach is to use regional "grid averages." Average analysis provides a very simple method to estimate system benefits of PV [1]. Given the time-varying nature of both PV output and power plant operation, "marginal" analysis provides a greater degree of accuracy when determining emissions or fuel displacement.

There are two general methods to marginal grid analysis that can be generally classified as "accounting" and "modeling."[2] Accounting methods attempt to collect historical generation information to estimate those units that are likely to reduce generation in response to the output from a renewable source such as PV. There are a number of advantages to this approach, one of which is a fairly realistic reflection of the current grid, and current grid operations strategies. Data sets used include estimates from individual utilities, various historical plant-level data sets, and more recently, the EPA continuous emissions monitoring system (CEMS) databases. Accounting methods have been previously used to estimate the impacts of limited deployment of PV [3-6].

Among the most significant limitations of accounting methods is the limited ability to "redispatch" the system based on changes in the generation mix due to the introduction of new generation technologies, including more than a relatively small amount of renewable energy generation. The use of simulation models allows for system re dispatch, and also allows for greater examination of the use of transmission. Models also allow for the dispatch of hydro resources, which may be important when simulating relatively large penetration of intermittent renewables.

3.0. PROJECT APPROACH AND METHODS

The approach of this study is to simulate the operation of electric power systems using a utility power plant dispatch model. Power plant dispatch is based on the actual operating (variable) cost of generation, including both fuel and operation and maintenance. Plants are dispatched from lowest to highest cost, based on the load, plant availability, and a variety of system constraints, such as power plant start-up times, ramp rates, environmental restrictions, transmission congestion, etc. Figure 1 illustrates an example dispatch scenario. The power plants dispatched first are those with the lowest variable costs, including nuclear, geothermal, and wind units. Some of these generation types, such as wind, have essentially zero variable cost, and are not controllable. Others, like nuclear, have some small fuel cost, but are difficult to ramp. Coal units typically have the next lowest cost, followed by combined-cycle (CC) and single-cycle gas turbines (CT).

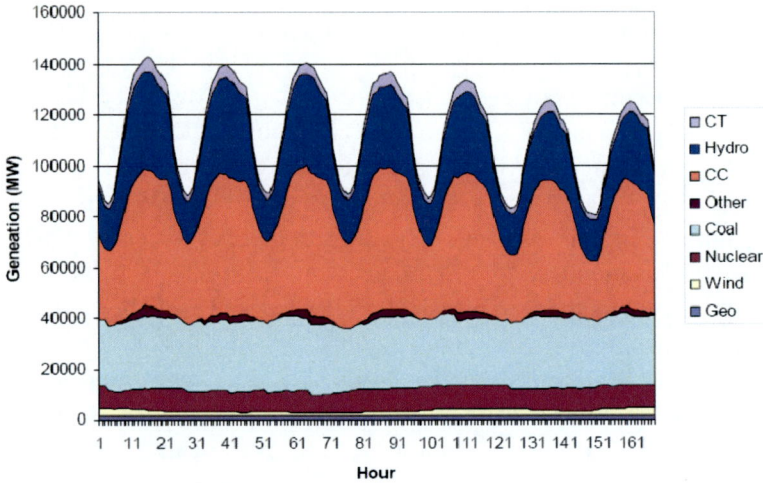

Figure 1. Representative System Dispatch for a Summer Week.

As can be observed, hydro dispatch is performed in a somewhat different manner from conventional thermal plants. It has essentially zero fuel cost, but also has limited energy availability. Hydro units also have the ability to ramp very quickly in response to variation in load.[1] Hydro is therefore typically dispatched as a load following and peaking plant, while operating under various environmental, recreation, and regulatory constraints of minimum and maximum water flows.

During real-time operations, increased load results in an increase in generation from the least cost unit available, while any reduction in the system load will result in the highest cost unit being "backed off." The marginal or incremental unit(s) vary from hour to hour. As can be seen in figure 1, any decrease in mid-day electric demand will affect primarily the CC units. Only after a substantial load reduction would there be any effect on coal units.

Utility system operators use a number of tools that estimate the most optimal dispatch of individual generators. These tools are referred to by several names, including "production cost," "unit commitment and dispatch," or "chronological dispatch" models. A high quality production cost model takes into account not only the variable cost of operating each plant, but also the large number of generator and system constraints to solve the optimal dispatch of all power plants in a utility fleet or an entire region. These constraints include several that may be very important when evaluating the impacts of PV.

Each power plant has operational limits, including the ability to ramp, minimum up and down times, and minimum loading. At high penetration of PV, the ability of power plants to reduce output may limit the amount of PV that can be accepted into the grid [7]. In addition to operational limits, each power plant has an efficiency or heat rate (fuel used per unit of generation) that varies as a function of load. As PV penetration increases, power plants may need to cycle more, resulting in lower average efficiency. This cycling could reduce the average fuel use and emissions offset as a function of PV penetration. (It will also increase the average cost of generation from thermal units, along with maintenance requirements. While

---

[1] Assumes hydro with dam storage, not "run-of-river" type plants.

the integration cost impacts of PV are an important consideration, they were not analyzed in this study.)

It should also be noted that while operational limits at the generator level are considered, there may be limits of PV deployment within the distribution system. These limitations are discussed in detail in several of the other Renewable Systems Interconnection studies.

For this study, we evaluated the optimal dispatch of power plants in several regions of the United States with and without PV. This evaluation consisted of performing a "base" run in each region without PV (0% PV penetration), then adding PV using simulated output from a distributed PV network.

The tool used for this study is PROSYM, offered by Global Energy Decisions. The tool comes with a database of the U.S. generation fleet, including heat rate curves and such constraints as minimum loading levels, along with a "reduced form" approximation of the transmission system. Accounting for transmission is one of the significant challenges in modeling electric power systems. The interconnected nature of the U.S. grid, and the power exchanges that occur over large regions must be considered when attempting to optimally dispatch the system as a whole.

3.1. Simulation of an Interconnected System

The electric power system in the United States consists of three large grids: the Eastern Interconnect, Western Interconnect (also known as the Western Electricity Coordinating Council or WECC), and the ERCOT (Texas) grid. All generators in each interconnect are synchronized and power may flow from any point to another within each grid, assuming transmission availability.

The use of transmission within each grid allows for a more reliable and cost-optimal system as a whole. Utilities typically contract for power and energy from other regions through a variety of open market and bilateral contracts, within the constraints of generation and transmission availability. This interconnectedness provides challenges when simulating the grid in any particular region. While utilities in certain areas may have sufficient generation to meet their load, it may be far more efficient for those utilities to purchase energy from a utility in a different region than run their own generation.

This study uses a "centralized dispatch" approach to system operation. PROSYM evaluates the system as a whole, dispatching all generators to optimize for least cost performance. This assumption is based in part on the existing levels of communication and cooperation that exist today, even though WECC is not centrally dispatched. Furthermore, it will be some time before PV achieves the high level of penetration evaluated in this study, and the electric power system will change physically and operationally. While we do not necessarily assume that WECC as a whole will become part of a centralized dispatched system or a single market, it is likely that continuous improvements in communication of price signals, transmission availability, etc, allow for our centralized dispatch model to be a reasonable approximation of the future electric power system as a whole.

Figure 2 provides the topology for this study. Within PROSYM, the Western Interconnect (WECC) is divided into a number of transmission areas, each comprising a load and a number of generators [8]. Within each transmission area, load flows are essentially unconstrained. Transmission between regions is modeled with a reduced form approximation based on a rated link between each transmission area. Power may flow between transmission areas, limited by path ratings, and taking into account line losses.

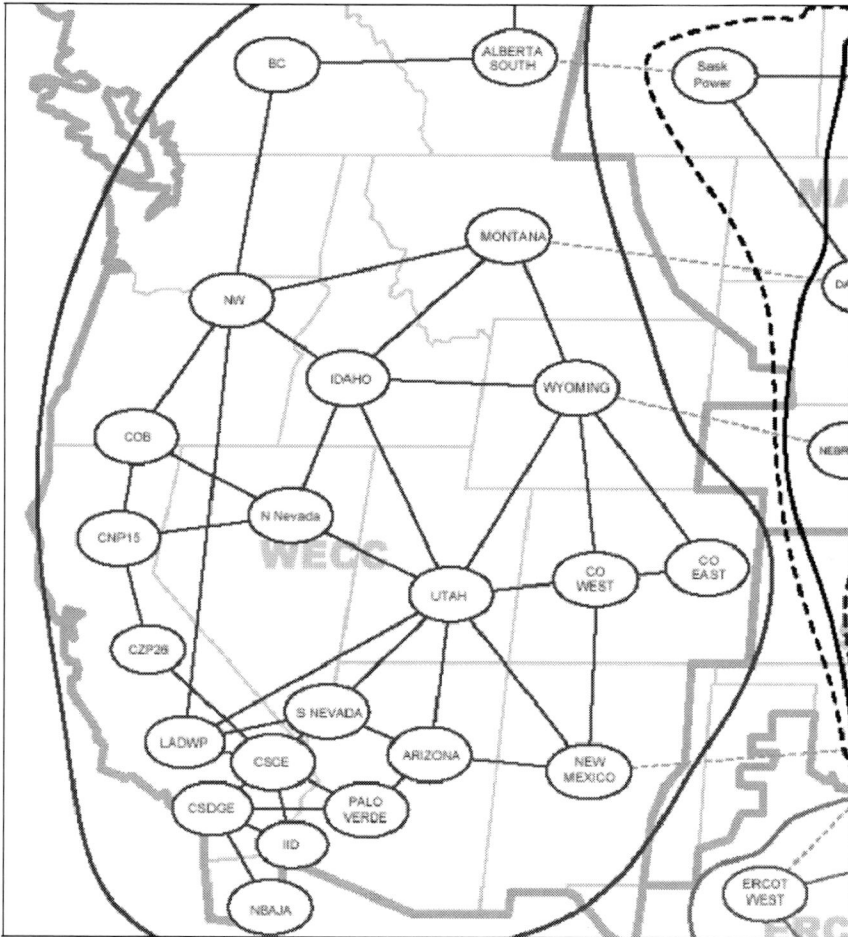

Figure 2. WECC System Topology Used by PROSYM.

For this study, we examined the impacts of PV on three aggregated regions – the state of California, consisting of seven transmission areas, the state of Colorado, consisting of two transmission areas, and the entire WECC region.

3.2. Assumed Scenario

Without generating a regional PV penetration scenario, it is not possible to capture the real power exchanges that will occur in an interconnected system. Therefore, it is important to create a scenario of PV deployment that considers interaction of local PV generation within the area of specific interest with the surrounding system. We generated a single overall scenario with PV deployment throughout the WECC region, while focusing on generator operation within California and Colorado. The scenario actually consists of a series of PV penetrations in which 1% to 10% of the entire western interconnect's annual electrical energy is derived from PV.

We began by obtaining hourly solar radiation data from the updated National Solar Radiation Database (NSRDB) [10], and simulating the performance of PV systems deployed at a variety of locations and orientations. A total of 75 sites within WECC were simulated, with each site having 14 possible configurations representing homes and buildings with various roof pitches and orientations, and also the use of utility tracking arrays. Since there is considerable correlation between system load, weather, and solar insolation, the solar data must match the load year. For this study, we chose 2003 as our base year for both insolation and load.

After the hourly solar output was simulated at each of the site and orientation combinations, a composite PV output was generated for each of the transmission areas modeled within PROSYM. This composite output was generated by weighing the contribution from each location based on its population. (We assumed that PV within a transmission area would be deployed roughly in proportion to local population, and we used Census data to match population with the distribution of PV.)

Once a composite hourly PV output was generated for each transmission area, an overall regional penetration scenario was developed. The base assumption is that PV will be built in states with the highest level of driving factors, including high electricity prices, incentives, political support, progressive utilities and rate structures, and good insolation.

Table 1 provides a list of the transmission areas that were assigned PV generation.

The majority of PV is assumed to be constructed in California (well over 50%). No PV was assigned to several regions, including the two Canadian provinces in WECC and the Northeastern part of WECC, including Wyoming, Eastern Idaho, and Montana.

Based on the geographical weighting of PV locations, the various overall penetration scenarios were created. Penetration scenarios were developed based on a 1%, 2%, 4%, 6%, 8%, and 10% penetration (by annual energy) of PV *in the entire WECC region*. It is important to consider this when evaluating the results of this study, particularly the results of the individual state analysis. In both the Colorado and California studies, the actual penetration of PV on an energy basis is higher than the named scenario. In the 10% scenario, PV is actually generating energy sufficient to meet 13.5% of Colorado's load, and 15.6% of California's load. Since scale factors were applied linearly, these adjustment factors can be applied to each of the named scenarios. (For example, in the 2% scenario, Colorado PV generation is equal to 2% * 1.135 and California PV generation is equal 2 * 1.156 etc.)

Model runs were performed for 2007, 2015, and 2020. Future loads are simple linear extractions based on estimated growth rates. It is important to note that the relative penetration of PV remains constant, so the only real change between the yearly simulations are changes in the regional generation mix. The generation mix for future years is built into the PRO SYM model, based on a "business as usual" scenario that includes certain state RPS policies, but no aggressive policies towards climate change. It is possible, however, to include such scenarios by altering the generation mix, or including carbon taxes or caps.

Table 1. Distribution of PV Generation

Transmission Area	Fraction of Total WECC Load (2007)	Fraction of WECC PV Capacity	Fraction of WECC PV Energy	Fraction of Region's Load Met by PV in the 10% Energy Scenario
Arizona	8.4%	10.0%	11.3%	14.7%
Northern California (N P26 + CZP26)	14.3%	22.2%	21.7%	14.2%
San Diego Gas & Electric	2.5%	4.0%	4.2%	17.1%
Southern Cal. Edison	13.2%	21.2%	22.2%	17.0%
Los Angeles Dept. of Water & Power	3.5%	5.6%	5.9%	14.1%
Imperial Irrigation District	0.4%	0.6%	0.7%	18.0%
Northern Nevada	1.5%	1.4%	1.5%	10.5%
Southern Nevada	3.4%	3.1%	3.5%	11.0%
Idaho Southwest	1.7%	1.2%	1.2%	6.2%
New Mexico	2.7%	3.2%	3.7%	15.0%
Utah	3.7%	3.9%	3.8%	9.5%
Northwest All of WA, OR, and far W. Mont.	17.7%	15.5%	11.9%	5.0%
Colorado West	0.7%	1.1%	1.0%	11.1%
Colorado East	5.5%	7.0%	7.4%	13.8%
Remainder of WECC	19.7%	0%	0%	0%

4.0. PROJECT RESULTS

To track various performance metrics, all generators in WECC were categorized into several groupings: combined-cycle gas turbines (CC), simple-cycle gas turbines (CT, in which we included gas-fired steam turbines and reciprocating engines to represent peaking plants), coal, nuclear, geothermal, hydro, pumped hydro storage, and wind. A relatively small number of plants not fitting these categories (mostly small thermal plants fired by a variety of fuels, including wood, waste, landfill gas, petroleum coke, etc.) were placed into an "other" category.

Simulation runs were performed for a base case (0% PV) and for each of the penetration scenarios. Hourly generation and fuel use was tracked from each power plant category, and emissions of carbon dioxide (CO_2) were tracked on a monthly basis. While one of the primary uses of production cost models is to track generation-related costs, these were not evaluated in this study.

4.1. Base System Characteristics

Base case runs (no additional PV) were performed with PROSYM to estimate the fuel mix for the current and future year scenarios. The results of the runs were also compared to historical data in an attempt to validate model assumptions.

Figure 3 indicates the WECC fuel mix for the study years, compared to actual data. The projected fuel mix changes slightly, with an increase in the fractional generation from gas and wind, and a decrease in fractional generation from coal.

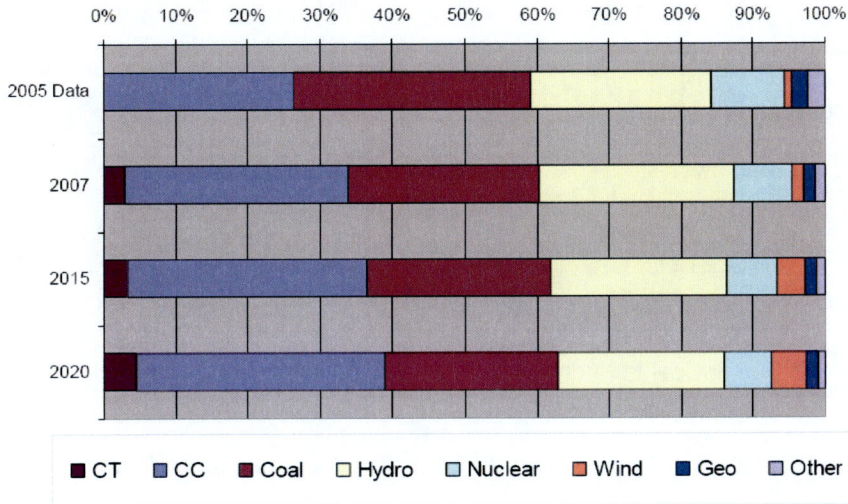

Figure 3. Historical Generation Mix and Simulated Generation Mix in WECC.

The most significant difference between the 2005 data and 2007 estimates is the fractional use of gas generation. (It should be noted that all gas-fired generation from the 2005 data, including combined- and simple-cycle gas turbines and gas steam units were included in the CC category). There are several possible explanations for this discrepancy. First, the amount of gas-fired generation has increased since 2005, accommodating virtually all the baseload growth in the demand. In addition, there are certain accounting differences in the "other" category for both the 2005 data and the 2007 model runs. In the 2007 simulations, the "CT" category actually includes all peaking plants, including those liquid-fueled steam turbines, and internal combustion engines. Some of these units are actually included in the "other" category in the 2005 data.

There are several other caveats regarding the comparison between the 2005 data and the future projections. The PROSYM simulations include British Columbia, Alberta, and Baja California, while the 2005 data includes only U.S. generation. These non-U.S. areas account for about 17% of the entire WECC load and may account for some of the differences. Finally, there is significant variation in hydro resource from year to year. Further data analysis is necessary to estimate the actual differences between historical data and model estimates, accounting for the differences in power plant accounting, nonU.S. generation, and hydro variation.[2]

Also compared was generation data for two states: California and Colorado. Figure 4 compares actual 2005 data with 2007 simulations for the state of California. As previously,

---

[2] A forthcoming version of this report will attempt to further reconcile the differences in historical data with simulations by comparing plant level performance and identifying any real differences. Also, the power plant data within PROSYM will be recategorized to isolate non-U.S. generators.

important caveats include variation in hydro availability, and accounting differences for a number of thermal generators using fuels other than coal and natural gas.

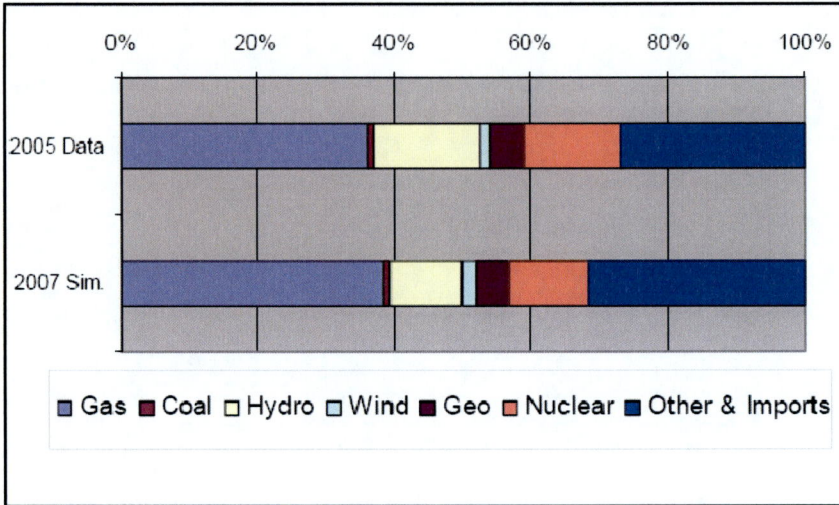

Figure 4. Historical Generation Mix and Simulated Generation Mix in California.

Figure 5 and Figure 6 compare an estimated actual plant dispatch in the California ISO from a summer day in 2006 [9] to a simulated dispatch in California in 2007. The simulated California dispatch includes the entire state, while the California ISO does not include several parts of northern and eastern California, and the Los Angeles Department of Water & Power, together accounting for about 12% of the state's load. Actual plant dispatch is difficult to compare because of how various plants are categorized. In the actual dispatch, both "thermal" stations and qualifying facilities include a large number of plant types, including CCs, geothermal, and industrial cogeneration plants (some of which may utilize CTs).

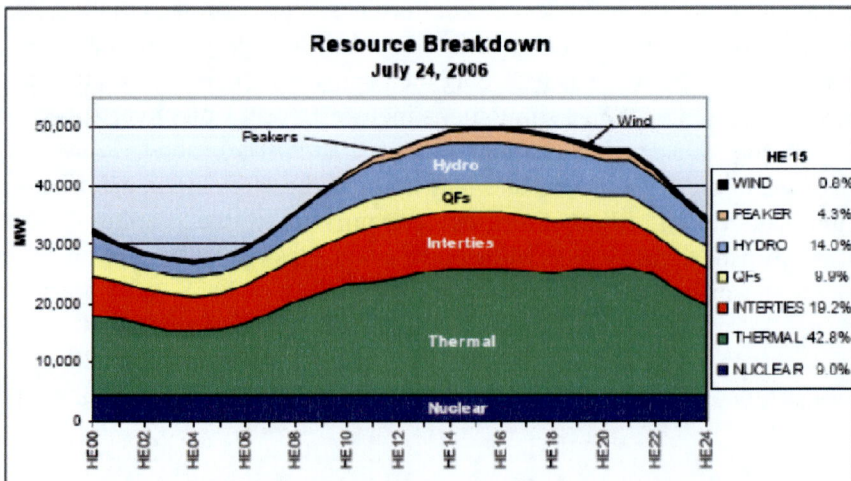

Figure 5. Historical Dispatch for CAL-ISO.

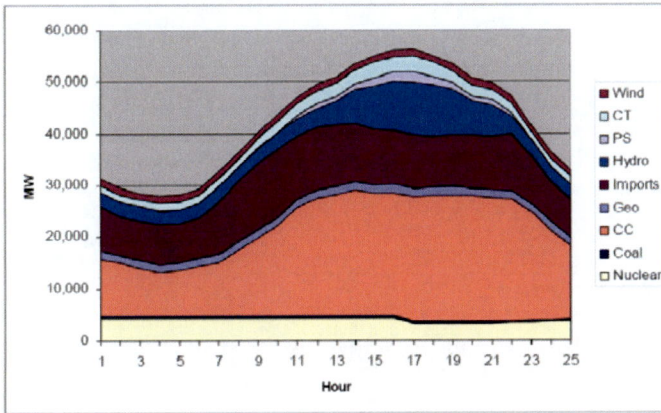

Figure 6. Simulated Dispatch for the State of California.

It is important to note that it is inappropriate to compare the actual plant dispatch to the simulated plant dispatch in any given hour, or over very short time periods. Variations in plant outages, wind availability, and various operational considerations make such a direct comparison of short-term data of limited value. Production cost model simulations may include both scheduled outages and random forced outages, or forced output reductions, which will not match "real" outages. As a result, this study is not intended to evaluate the impact of PV during a specific hour or day, but is intended to evaluate the longer-term impacts (seasonal to annual) of PV deployment.

4.2. Load Shape Impacts

Introduction of customer-sited PV will change the overall load and load shape met by conventional generation. The amount of load reduction and the time and season of load reduction will determine the mix of avoided generation.

Figure 7, Figure 8, and Figure 9 illustrate the type and magnitude of load shape impacts created by the various levels of PV penetration in each region. The 1% case is omitted for clarity. In each graph, three representative 2-day periods (summer, spring minimum, and summer maximum) are used to illustrate simulated PV impacts for the year 2007. During the winter, variation in electricity demand is driven largely by heating and lighting, with two daily peaks: a morning peak and a larger evening peak driven largely by lighting. Winter PV generation occurs in between these two peaks and will not reduce overall peak demand. Spring loads are fairly flat during the daytime given the minimal need for heating or air-conditioning, with a relatively small evening lighting peak, again unaffected by PV generation. The minimum demand for electricity generally occurs in the overnight hours in the spring season. Summertime peak loads are driven by air conditioning demand, which is largely coincident with PV output. As a result, PV can act to reduce peak demand, and will act to offset generation from potentially lower efficiency peaking plants, such as simple-cycle combustion turbines.

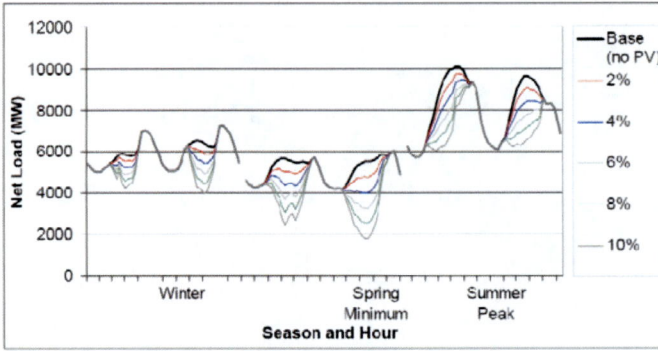

Figure 7. Load Shapes in Colorado with Various WECC PV Penetration Scenarios.

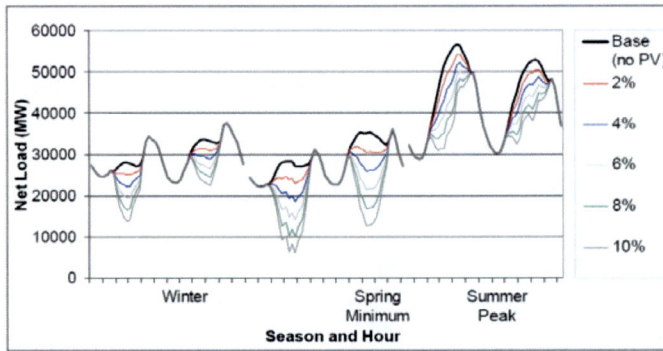

Figure 8. Load Shapes in California with Various WECC PV Penetration Scenarios.

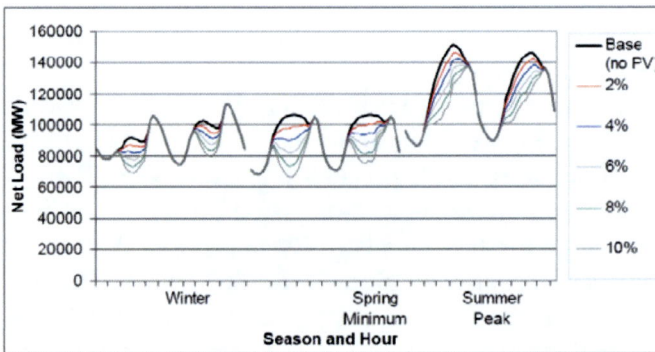

Figure 9. Load Shapes in WECC with Various PV Penetration Scenarios.

The overall load shapes in California, Colorado, and WECC as a whole (which includes both California and Colorado) are fairly similar. The net load shape with PV in WECC is considerably smoother than in the individual states. This is largely due to the aggregation of the 75 PV locations, while the net loads in individual states use fewer PV sites. In reality, the composite PV profile in a state will potentially be smoother due to the large number of distributed PV sites. While probably not a major influence of the outcome of this study, the more irregular PV profile might increase the ramping requirement of the system, and future studies should probably include many more sites within each transmission area.[3]

Overall general impacts on loads can be observed through the use of a Load Duration Curve (LDC). Figure 10 illustrates an LDC for the entire WECC region for several PV penetration scenarios in 2007. The load duration curve shapes for California and Colorado are quite similar, with only the magnitude of the load changing.

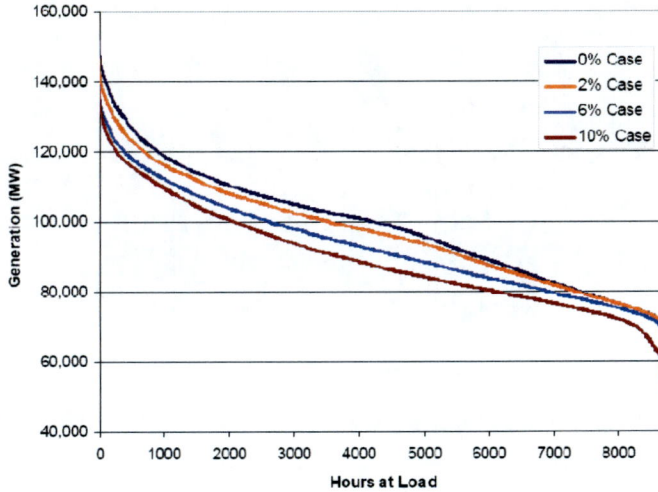

Figure 10. Load Shapes in WECC with Various PV Penetration Scenarios.

Among the more noticeable features in Figure 10 is the reduction in annual minimum load that occurs in high penetration. This implies that at high penetration, PV will begin to offset "baseload" generation [9].

Because the future year scenarios (2015 and 2020) simply grow the 2003 load, the load shape impacts of PV are identical. This assumes that there are no long-term changes in solar output due to climate, and that electricity usage patterns stay constant over time. Sensitivities to these assumptions may be evaluated in future analysis.

4.3. Avoided Generation

As previously discussed, PROSYM dispatches the entire Western Interconnect and optimally dispatches the entire power plant fleet. The generation in individual areas can be isolated to examine the changes in power plant dispatch. Generators of a common type in each of the study regions (California, Colorado, and WECC as a whole) were grouped to examine PV impacts on the various generator types. The net generation within a transmission area can also be compared to the load. This establishes the net import and export of electricity. While it is not possible to track the origin and destination of every unit of energy, looking at net imports is useful, especially when the remainder of the system can be characterized.

---

[3] One counter to this issue is the fact that only hourly data are used. Hourly data will tend to filter out such phenomena as passing clouds. However, production cost models are typically run on hourly intervals and may not capture some of the dynamic aspects of intra-hour variations of PV output.

4.3.1. Avoided Generation in California

Figure 11 and Figure 12 show simulated generation for California in a summer and winter day in 2007 for each PV penetration scenario (1% is omitted for clarity). In both cases, offset generation is primarily from combined-cycle generations, with some reduction in net imports at high penetration.

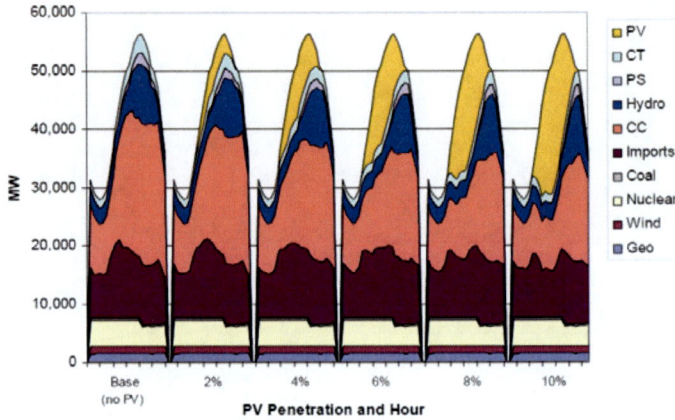

Figure 11. Simulated Dispatch in California for a Summer Day in 2007 with Various PV Penetration Scenarios.

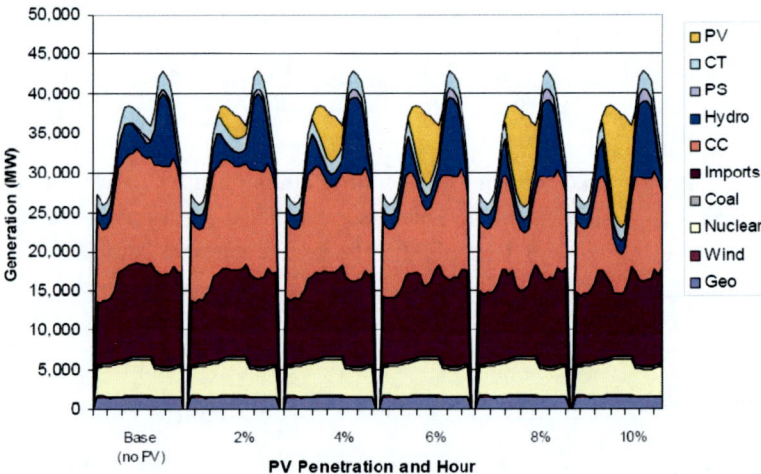

Figure 12. Simulated Dispatch in California for a Winter Day in 2007 with Various PV Penetration Scenarios.

The actual mix of displaced generation is illustrated in Figure 13 and Figure 14. Figure 13 describes the total mix of ALL displaced generation at various penetration levels in the 2007 case, dominated by natural gas-fired units.

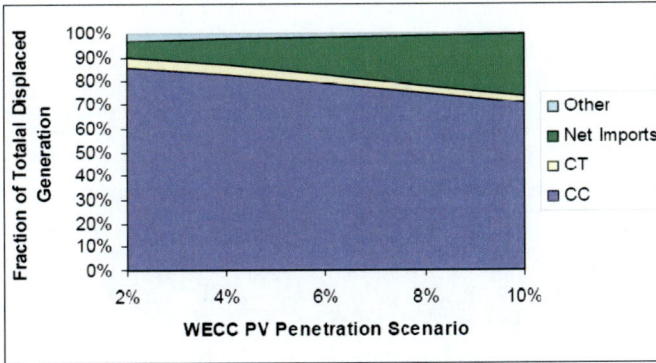

Figure 13. Mix of Displaced Generation from PV Deployed in California.

Figure 14 illustrates the incremental or marginal displaced generation in each "step" of PV installation. In the highest penetration case, going from 8% to 10% of all WECC generation from PV, nearly 50% of this incremental PV generation in California is offsetting generation outside the state of California.

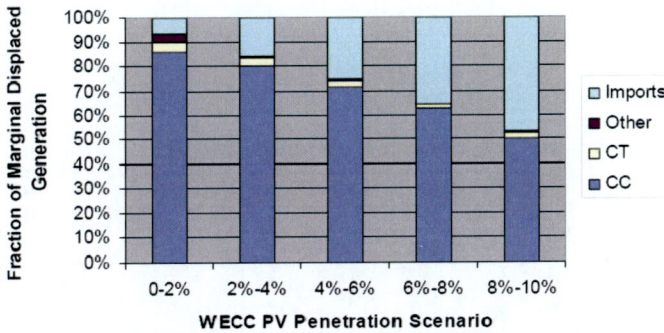

Figure 14. Mix of Incremental Displaced Generation from PV Deployed in California.

4.3.2. Avoided Generation in Colorado

Figure 15 and Figure 16 illustrate simulated dispatch scenarios for Colorado. Compared to California, Colorado imports a much lower fraction of its electricity, and also relies more heavily on coal.

Figure 15 illustrates a spring day, demonstrating the fact that Colorado meets most of its baseload demand from coal. Up to about the 4%to 6% scenario, PV displaces mostly CC and imports on this day. Beyond this point, PV begins to displace coal generation. During certain hours, imports are completely displaced, and the state becomes a net exporter of electricity. (While the graph implies that coal and wind are being exported, we are not explicitly tracking imports and exports at the plant level, and the origin of the exports cannot be explicitly identified.)

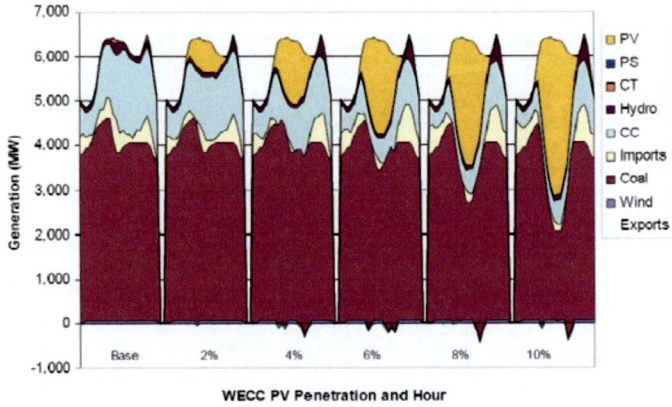

Figure 15. Simulated Dispatch in Colorado for a Spring Day in 2007 with Various PV Penetration Scenarios.

Figure 16 provides the results of a summer day simulation. The greater baseload demand results in even less coal displacement, and most PV generation displaces natural gas-fired generators. As before, the area of negative generation represents periods where there is a net export (imports are negative) of electricity from the state.

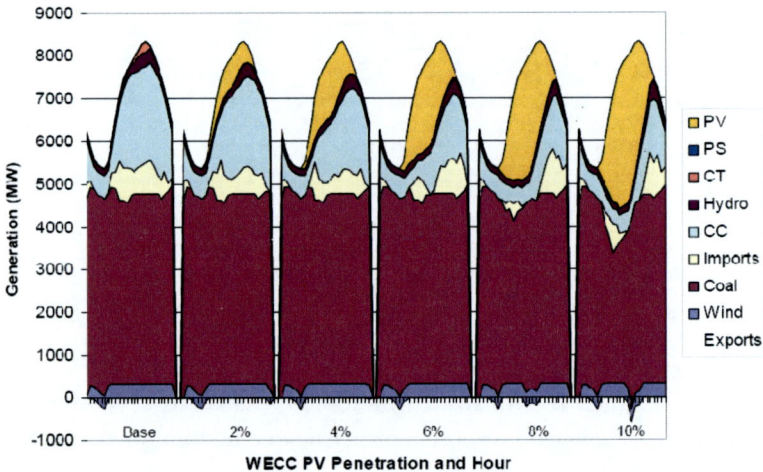

Figure 16. Simulated Dispatch in Colorado for a Summer Day in 2007 with Various PV Penetration Scenarios.

Figure 17 illustrates the total fractional mix of displaced generation.

Figure 18 illustrates the incremental fractional mix of displaced generation. In the 8% to 10% WECC penetration scenario, about 60% of this incremental PV generation in Colorado is offsetting coal-fired generation.

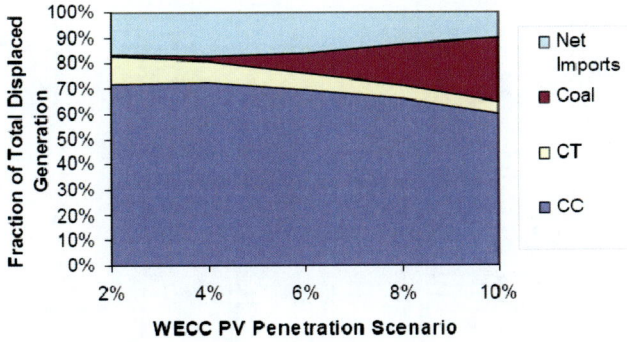

Figure 17. Mix of Total Displaced Generation from PV Deployed in Colorado.

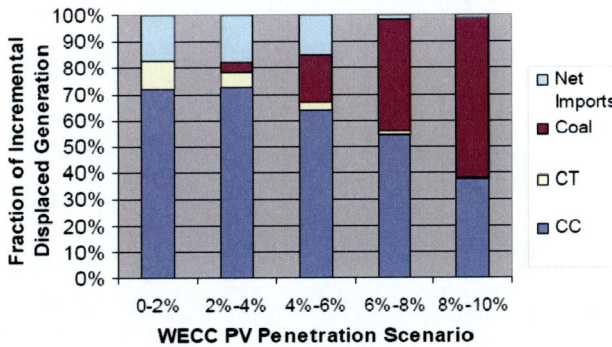

Figure 18. Mix of Incremental Displaced Generation from PV Deployed in Colorado.

4.3.3. Avoided Generation in WECC

Figure 19, Figure 20, and Figure 21 illustrate the representative impacts over the entire WECC Region (including California and Colorado) for representative winter, spring, and summer days.

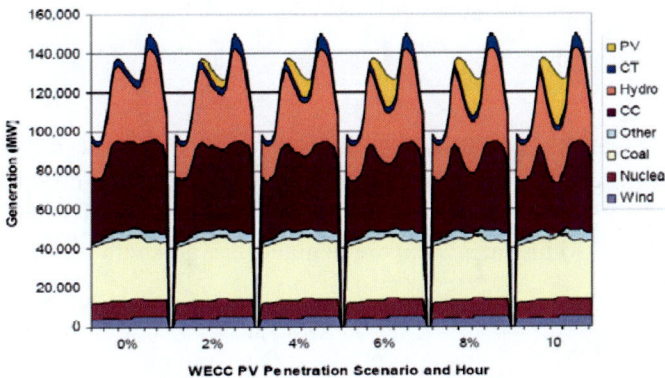

Figure 19. Simulated Dispatch in WECC for a Winter Day in 2007 with Various PV Penetration Scenarios.

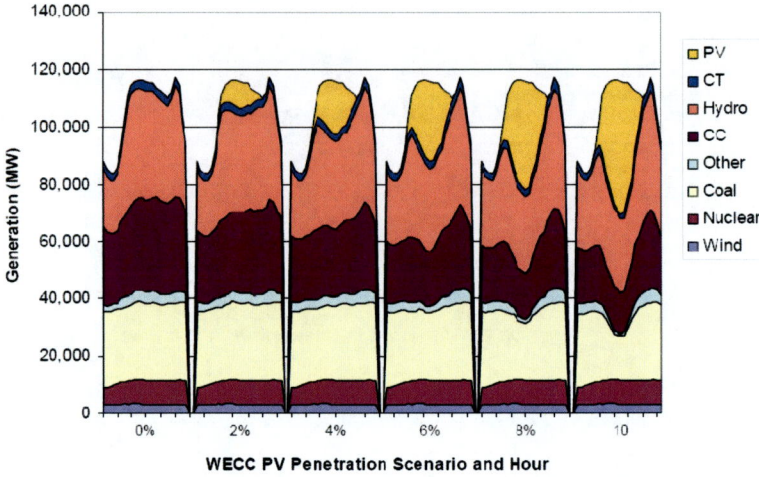

Figure 20. Simulated Dispatch in WECC for a Spring Day in 2007 with Various PV Penetration
Scenarios.

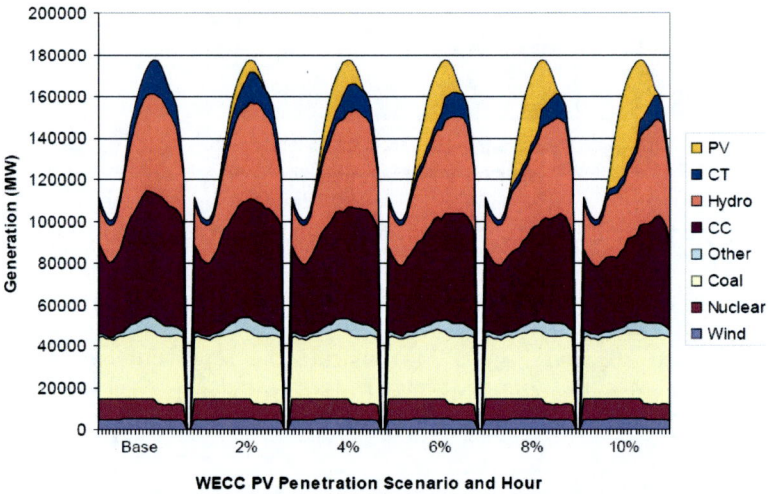

Figure 21. Simulated Dispatch in WECC for a Summer Day in 2007 with Various PV Penetration
Scenarios.

The previous simulations indicate that taken as a whole, the assumed mix of PV locations
results in mostly displacement of gas-fired generators. Figure 22 illustrates that at a 10%
penetration, more than 85% of the total expected offset generation will occur from natural gas-
fired generators. Figure 23 illustrates the incremental offset generation.

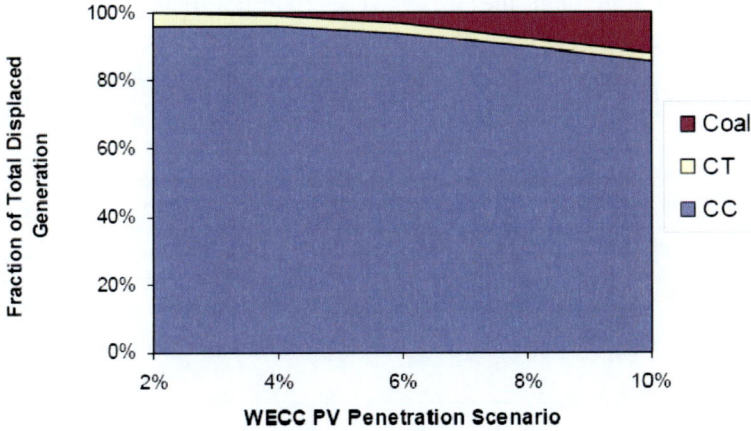

Figure 22. Mix of Total Displaced Generation from PV Deployed in WECC.

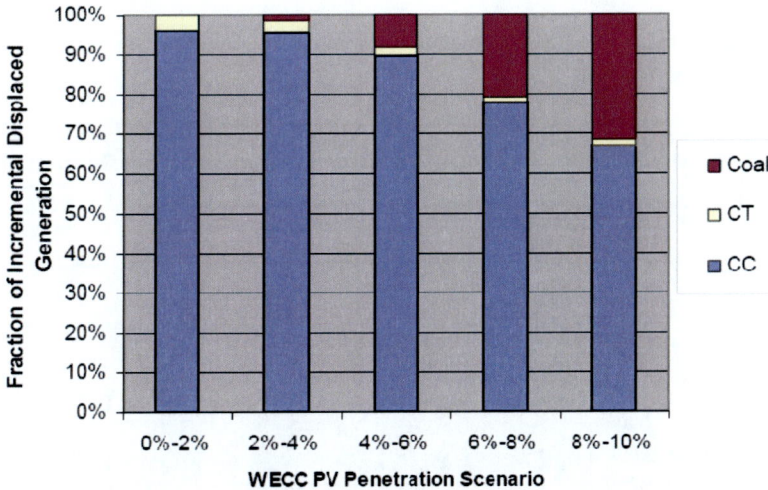

Figure 23. Mix of Incremental Displaced Generation from PV Deployed in WECC.

4.4. Avoided Fuel Use

The avoided generation estimates can be translated into avoided fuel, and produce a "fuel content" for a kilowatt-hour (kWh) of electricity generated by a PV system in various regions within WECC. In addition to the variation in generator types, the model simulates the effect of part-load operation. If PV increases the amount of power plant cycling, this may result in higher average heat rates for plants following the variation in output from distributed PV and a corresponding decrease in offset emissions rates.

4.4.1. Avoided Fuel Use in California
Figure 24 illustrates the average gas displacement rate for PV generation in the state of California. Three lines are shown: the displacement rate for PV when offsetting CT

generation, CC generation, and the weighted average of both (dominated by CCs as demonstrated in Figure 13). It is important to note that this offset applies only to the fraction of generation that effectively "stays" in California. These results can be combined with the fraction of in-state generation offset by PV in Figure 13.

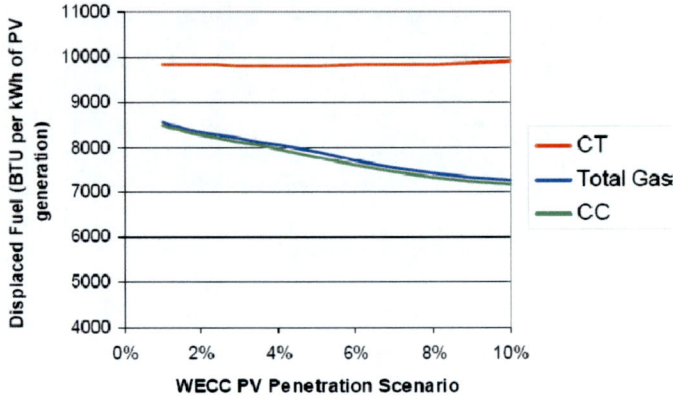

Figure 24. Average Natural Gas Fuel Displacement from PV Deployed in California and Offsetting California Generation.

Figure 25 illustrates the marginal displacement rate for California generation. As before, this only applies to the fraction of PV generation that displaces in-state generation as estimated in Figure 13.

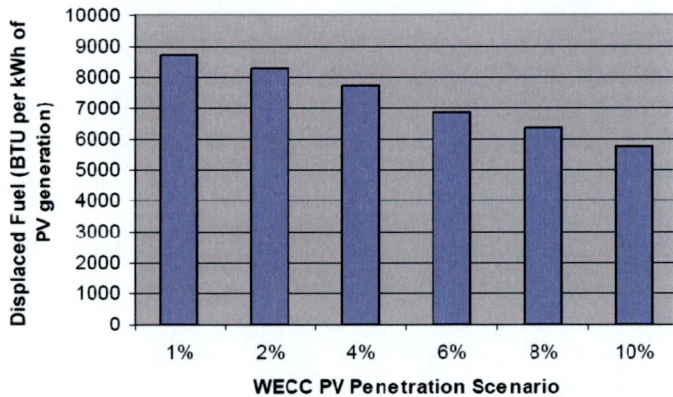

Figure 25. Incremental Natural Gas Fuel Displacement from PV Deployed in California and Offsetting California Generation.

The decrease in fuel benefits illustrated in Figure 24 and Figure 25 shows not only the increased displacement of more efficient generators as a function of penetration, but also the impacts of increased cycling. Figure 26 demonstrates the overall increase in gas unit heat rates that result from the increased cycling.

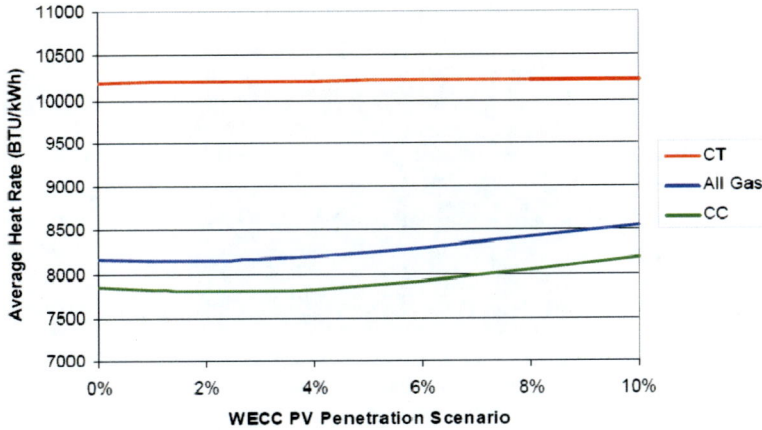

Figure 26. Average Heat Rates of California Natural Gas Generators Resulting from PV Load Following.

4.4.2. Avoided Fuel Use in Colorado

Within Colorado, both natural gas and coal is displaced by PV. Figure 27 illustrates the fuel offset rate for each of the plant types displaced by PV.

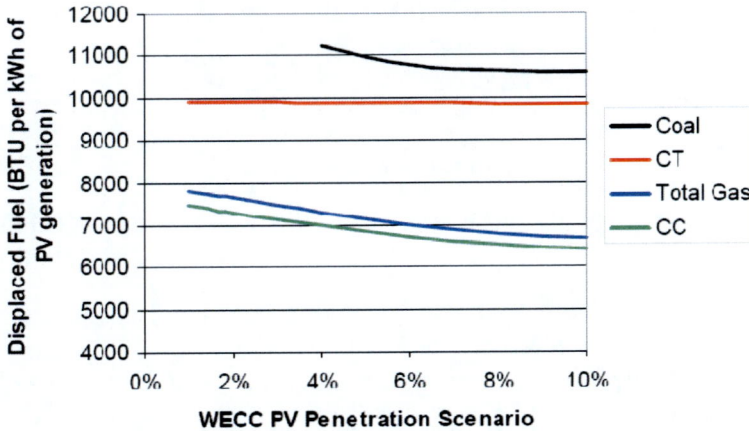

Figure 27. Average Fuel Displacement Rates from PV Deployed in Colorado and Offsetting Colorado Generation.

Using the estimated displacement mix of in-state generation from Figure 17 it is possible to estimate the overall average fuel displacement from 1 kWh of PV generation used within Colorado. Figure 28 illustrates the average total fuel displacement from in-state PV generation, while Figure 29 illustrates incremental fuel displacement.

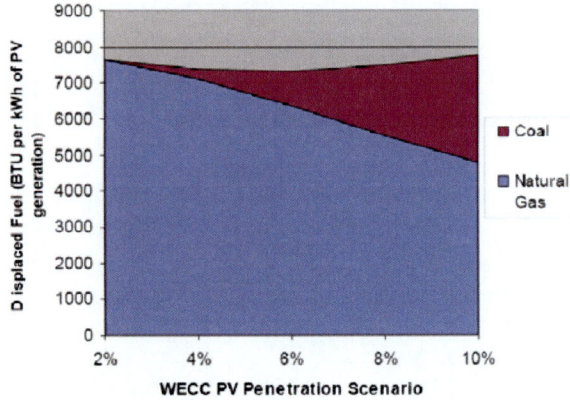

Figure 28. Total Average Fuel Displacement from PV Deployed in Colorado and Offsetting Colorado Generation.

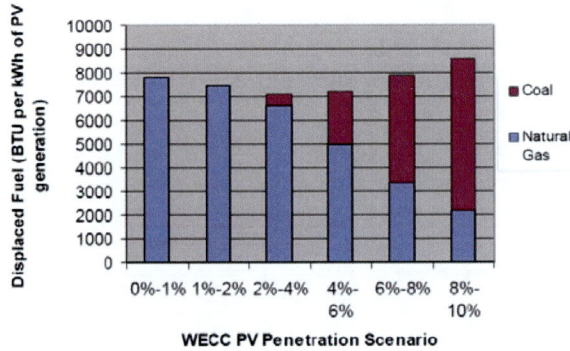

Figure 29. Incremental Fuel Displacement from PV Deployed in Colorado and Offsetting Colorado Generation.

4.4.3. Avoided Fuel Use in WECC

The overall fuel displacement rate within the entire WECC region is illustrated in figure 30 and figure 31.

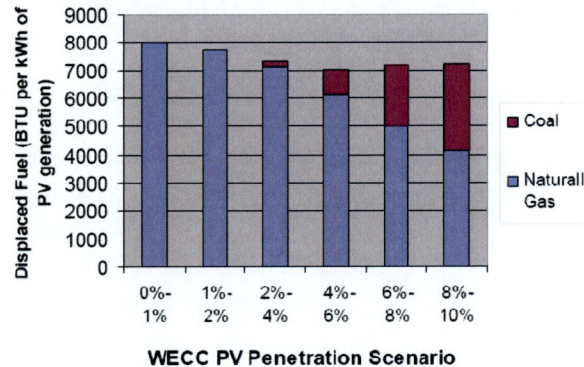

Figure 30. Total Average Fuel Displacement from PV Deployed in WECC.

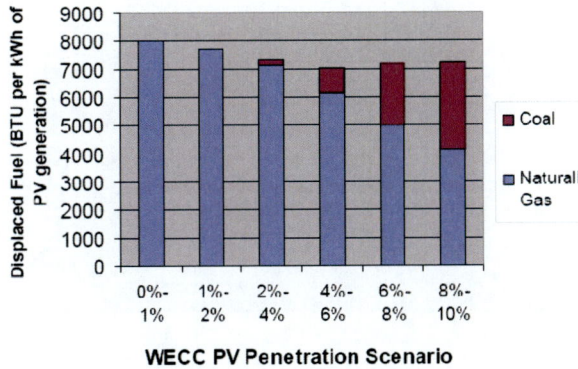

Figure 31. Incremental Fuel Displacement from PV Deployed in WECC.

4.5. Avoided Emissions

Estimates were produced for the avoided emissions of CO_2, NO_x and SO_2.

4.5.1. Avoided Emissions in California

Figure 32 illustrates the CO_2 emissions offset rate in California. Both marginal and incremental offset rates are shown. The decrease in emissions benefits as PV penetration increases is due to both the reduced displacement of less efficient generators, and increased fuel use associated with power plant cycling. As before, these rates apply only to the portion of PV generation that offsets California generation.

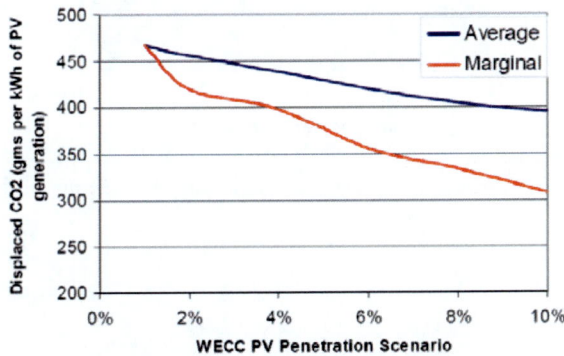

Figure 32. Average and Marginal CO_2 Emissions Displacement from PV Deployed in California and Offsetting California Generation.

SO_2 emissions are primarily associated with coal combustion. Because there is very little coal-based electricity generation in California, only NO_x emissions were evaluated. Figure 33 estimates the NO_x offset rate for PV generation that reduces in-state generation. There is initially a small decrease in the NO_x offset rates as PV displaces more efficient units, then an increase resulting from the offset of oil-fired units with higher NOx emission rates.

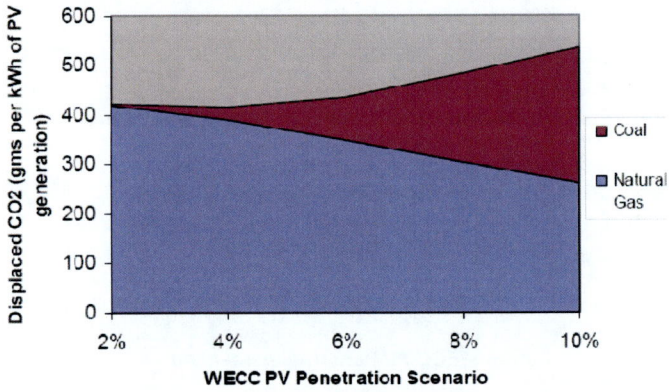

Figure 33. Average and Marginal NOx Emissions Displacement from PV Deployed in California and Offsetting California Generation.

4.5.2. Avoided Emissions in Colorado

Figure 34 and Figure 35 illustrate the average and marginal CO_2 emissions offset rates, showing the mix of avoided emissions from both coal and natural gas plants.

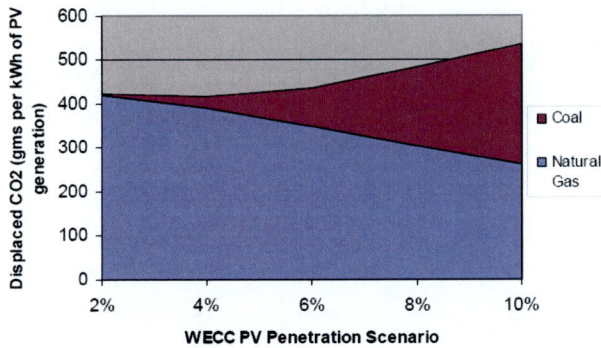

Figure 34. Total Average CO_2 Emissions Displacement from PV Deployed in Colorado and Offsetting Colorado Generation.

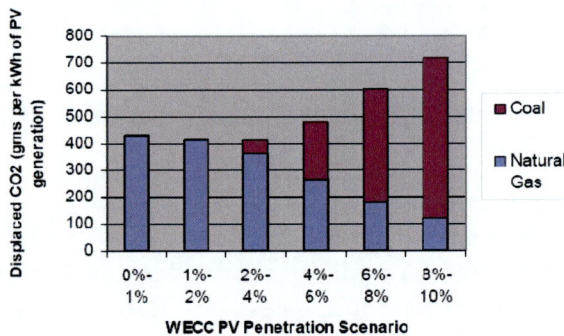

Figure 35.   Incremental CO_2 Emissions Displacement from PV Deployed in Colorado
and Offsetting   Colorado Generation.

The   estimated NO_x and SO2 offset rates are provided in Figure 36,
demonstrating   the greater emissions rates associated with coal-fired generation.

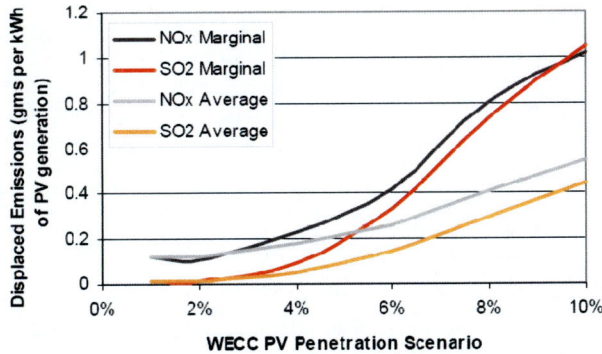

Figure 36. Average and Marginal NO_x and SO2 Emissions Displacement from PV Deployed in
Colorado and Offsetting Colorado Generation.

4.5.3. Avoided Emissions in WECC

The overall CO2 emissions displacement is driven by the fuel displacement values
illustrated in Figure 30 and Figure 31. The average and marginal values are provided in
Figure 37.

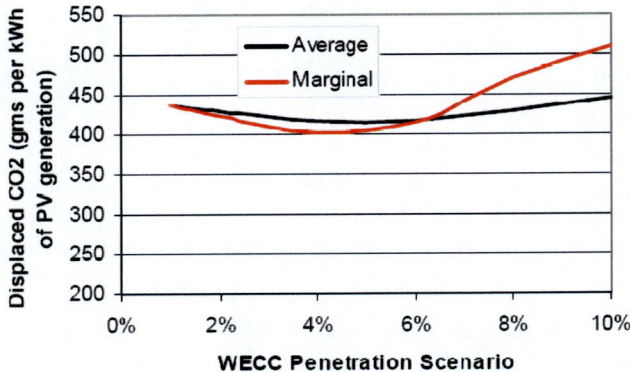

Figure 37. Average and Marginal CO_2 Emissions Displacement from PV Deployed in WECC.

Overall, there is a substantial variation in emissions displacement on a seasonal basis.
Figure 38 illustrates the incremental CO2 emissions displacement for various penetration

scenarios in each month. At low penetration, PV offsets high emissions peaking units during the summer, and more efficient combined-cycle units during the off-peak seasons. The incremental emissions rates then drop as PV starts offsetting more efficient units during all seasons. At higher penetrations (above 2% to 4%) PV starts to offset coal units, and displaced emissions rates increase.

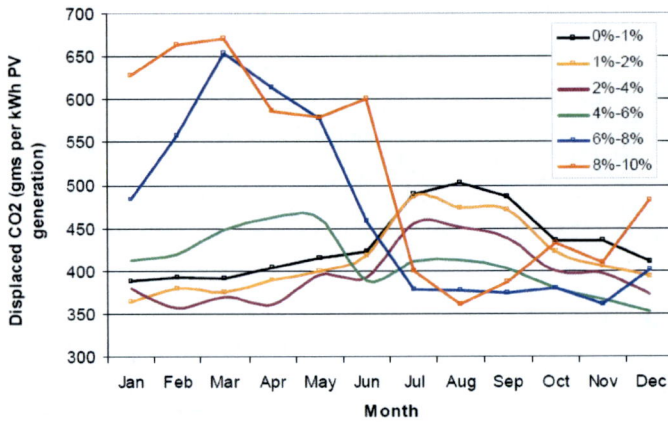

Figure 38. Seasonal Incremental CO_2 Emissions Displacement from PV Deployed in WECC.

Figure 39 illustrates the estimated offset rates for NO_x and SO_2 for the entire WECC scenario.

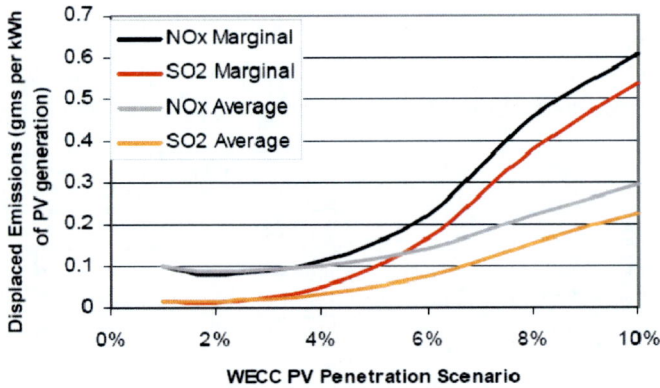

Figure 39. Average and Marginal NO_x and SO_2 Emissions Displacement from PV Deployed in WECC.

The seasonal patterns of emissions reductions for NO_x and SO_2 are similar to those for CO_2 (as illustrated in Figure 38), because the largest impact on coal generation from PV occurs in the spring and early summer.

4.6. Future Scenarios

The results presented in sections 4.3 through 4.5 represent the penetration of PV into the existing grid. Scenarios were also examined using Global Energy's projections of the generation mix in 2015 and 2020. Many of the changes in the grid expected to occur in this time frame, such as the installation of new baseload generation (wind, coal, or geothermal) will have little impact on the marginal generation affected by PV. As a result, the future scenarios are generally similar to the results presented for 2007. Among the most significant differences between the 2007 grid and the 2015 and 2020 grids, as projected in the model, is the greater overall reliance on natural gas, both in combined- cycle and simple-cycle gas turbines, illustrated in Figure 1. In the 2015 and 2020 simulations, this reliance "delays" the offset of coal generation until higher PV penetration is achieved. In addition, the projected future grid also relies more heavily on simple-cycle units. If true, this greater use of less efficient generators would increase the overall benefits of PV generation.

Given the fact that it will take some time for PV to reach the levels of penetration evaluated in this work, the future mix of generator types and their operation in response to intermittent generators, are important considerations. Follow on studies will evaluate a variety of capacity expansion scenarios

5.0. RECOMMENDATION FOR FUTURE RESEARCH

Recent analysis and evaluation of wind integration may provide both "lessons learned" and a general path forward for continued evaluation of the impacts of large scale PV deployment into the grid. The analytic questions utilities and the wind industry have been addressing over the last 10 years are similar to the questions that will have to be addressed for PV as this technology penetrates the market. The types of tools and analysis used in wind integration studies are similar to those used in this study; and the solar industry can benefit from the methods that have been developed to understand the impact of stochastic energy resources in electric power systems.

There are many opportunities for research into grid-level impacts of PV. (It should be noted that this list applies only to grid impacts at the generator level. It does not consider any aspects of the many "distributed" benefits or impacts of PV). Important issues for future research include:

- *Solar forecasting and Unit Commitment.* This study assumes prior knowledge of both load and solar resource. It is unclear how accurately utilities will be able to predict their net load with PV for day-ahead and hour-ahead unit commitment. The cost impacts of forecasting errors and uncertainty on utility operations should be explicitly examined.
- *Hydro Dispatchability.* The new and different load shapes created by PV deployment will require examination of the capacity of hydro resource to be dispatched to these new patterns.
- *Capacity Credit* – While there have been a number of analyses of the "capacity credit" of PV, there is significant additional work to be done, especially using the variety of

metrics used by individual utilities and system operators. Furthermore, much of the capacity credit analysis has occurred at the hourly time scale. This scale may be too long for utilities to have high levels of confidence in the ability of PV to serve load during peak demand periods. Other potentially important questions related to this topic include: How does the capacity credit change as a function of penetration? How can capacity credit be increased, considering system orientation and spatial diversity? Is the data quality and quantity sufficient to derive dependable capacity credit metrics?

- *Peak Demand Day Analysis.* PV could be a useful tool for improving air quality on peak demand days. Very detailed examination of PV impacts during theses days should be examined, possibly including sub-hourly analysis to capture actual impacts on peaking generators.

- *Combined Technology Studies.* It is very important to examine the system impacts of multiple renewable technologies including wind, concentrating solar power, and PV.

- *Sub-hourly Impacts.* What are the effects of sub-hourly PV ramping?

- *Incorporating T&D losses.* Tools such as PROSYM treat the load at the busbar, and do not consider how variations in load affect T&D losses.

- *Intermittency mitigation techniques* Previous wind integration studies have found modest costs at penetrations beyond 20% (on an energy basis) given sufficient spatial diversity, forecasting ability, and the ability to schedule and commit conventional energy resources over large areas. It is not clear at what levels of penetration PV will be burdened by "excessive" integration costs. Assuming such a level does exist, it may be important to examine enabling technologies and techniques, including increased spatial diversity; diversity of orientation; market-based approaches, such as time-of-use and real-time pricing; and technology options, such as load shifting, long distance transmission, and various centralized and distributed energy storage technologies. Of particular interest may be the use of plug-in hybrid electric vehicles as a PV enabling technology.

- *Limitations of Existing Tools.* The existing suite of utility simulation tools were designed to examine operations of conventional power stations. In most cases intermittent renewables have been "retrofitted" and there are still limitations in the treatment of technologies such as solar and wind. For example, PROSYM uses a single time zone for the entire WECC region, which may introduce errors when scheduling power flows from PV across the two WECC time zones. Treatment of hydro dispatch and coordination of hydro and thermal generations may also need improvement.

- *High Penetration Impacts.* In the simulated scenarios, the overall 10% penetration case created much higher penetration in certain regions in California. The net loads in these regions dropped to a very small fraction of the normal load, which may very well "push the limits" of the model's capabilities. High penetration scenarios require a greater understanding of system boundary conditions, including minimum load levels on existing plants, and hydro limitations. In addition, more transmission load flow studies will be needed to verify the system capabilities assumed in this analysis.

- *Sensitivity to PV Location.* This study assumes a fixed set of regional penetrations of PV based on existing policies and population patterns. These assumptions are conjectural and it may be useful to examine sensitivity of the results to a variety of PV deployment patterns.

- *Impact of Electricity Use Pattern.* This study assumes that future electricity use patterns remain the same. This may be unrealistic, given the increased use of time-of-use rates, and the possible use of real-time pricing, both of which could alter load patterns.

CONCLUSION

The use of production cost models allows for the estimation of the system impacts of large-scale deployment of PV. Based on a PV deployment scenario in the western United States where PV is mostly utilized in the Southwest and California, the following conclusions are generated:

- At low penetration (less than 4%), virtually all PV offsets generation from natural gas-fired units, primarily high-efficiency combined-cycle units.
- The natural gas fuel and emissions displacement rate for PV falls as a function of penetration as PV begins to displace more efficient gas units, and also creates increased plant ramping and part load operation.
- Increased penetration of PV (above 4%) results in greater levels of displaced coal generation, primarily in the high solar output months and low demand period in the late spring.
- At the highest penetration evaluated (10%) natural gas provides the majority of fuel offset, although the coal offset rate is rising rapidly.

Up to the 10% penetration case, the net load shapes created by PV appear to fall well within the operational capabilities of the regional grid and the PROSYM model. During a few hours of the year (mid-day in late spring), the net loads created by PV have fallen well below normal load conditions. In this case, PROSYM begins to see conditions close to "minimum load" levels that might require PV curtailment. However, significant additional work is needed to evaluate how well PROSYM characterizes operation of the power system at these very low load levels.

Given the relatively immature state of analysis of the effects of large-scale deployment of PV on the grid, it is recommended that continued efforts be made to develop appropriate data sets, analysis tools, and techniques. Lessons learned from the wind industry and the tools and methods developed for wind analysis will provide a useful start to this process.

REFERENCES

[1] Keoleian; Lewis. "Modeling the Life Cycle Energy and Environmental Performance of Amorphous Silicon BIPV Roofing in the U.S.," *Renewable Energy*, Vol. 28, 2003; pp. 271-293.

[2] Berlinski, M. *Quantifying Emissions Reductions from New England Offshore Wind Energy Resources*. Cambridge, MA: M.S. Thesis. Massachusetts Institute of Technology, 2006.

[3] Spiegel, R.J.; Edward, J.; Kern, C.; Greenberg, D.L. "Demonstration of the environmental and demand side management benefits of grid-connected photovoltaic power systems." *Solar Energy*; Vol.62, No.5; pp 345–58.

[4] Spiegel, R.J.; Greenberg, D.L.; Kern, E.C.; House, D.E. "Emissions Reduction Data for Grid-Connected Photovoltaic Power Systems."*Solar Energy;* Vol. 68, No. 5, 2000; pp. 475–485..

[5] Spiegel, R.J.; Leadbetter; M.R.; Chamu, F. "Distributed grid-connected photovoltaic power system emission offset assessment: statistical test of simulated- and measured- based data." *Solar Energy;* Vol. 78 (2005) 717–726

[6] Connors, S.; Martin, K.; Adams, M.; Kern, E.; Asiamah-Adjei, B. *Emissions Reductions from Solar Photovoltaic (PV) Systems. LFEE Report No.: 2004-003 RP* August 200 4.

[7] Denholm, P.; Margolis, R.M. "Evaluating the Limits of Solar Photovoltaics (PV) in Traditional Electric Power Systems." *Energy Policy*; Vol.35, 2007; pp. 2852-2861.

[8] Global Energy Decisions. *PROSYM User Guide*. Software Version 5.5 June 2007

[9] California Independent System Operator. *California ISO 2007 Summer Loads and Resources Operations Assessment*, March 2007

[10] National Renewable Energy Laboratory, National Solar Radiation Database 1991– 2005 Update: User's Manual NREL/TP-581-41364, http://rredc.nrel.gov/solar/old_data/nsrdb/1991-2005/.

In: Renewable Energy Grid Integration   ISBN: 978-1-60741-324-0
Editor: Marco H. Balderas   © 2009 Nova Science Publishers, Inc.

Chapter 3

PHOTOVOLTAICS BUSINESS MODELS*

L. Frantzis, S. Graham,
R. Katofsky and H. Sawyer

PREFACE

Now is the time to plan for the integration of significant quantities of distributed renewable energy into the electricity grid. Concerns about climate change, the adoption of state-level renewable portfolio standards and incentives, and accelerated cost reductions are driving steep growth in U.S. renewable energy technologies. The number of distributed solar photovoltaic (PV) installations, in particular, is growing rapidly. As distributed PV and other renewable energy technologies mature, they can provide a significant share of our nation's electricity demand. However, as their market share grows, concerns about potential impacts on the stability and operation of the electricity grid may create barriers to their future expansion.

To facilitate more extensive adoption of renewable distributed electric generation, the U.S. Department of Energy launched the Renewable Systems Interconnection (RSI) study during the spring of 2007. This study addresses the technical and analytical challenges that must be addressed to enable high penetration levels of distributed renewable energy technologies. Because integration-related issues at the distribution system are likely to emerge first for PV technology, the RSI study focuses on this area. A key goal of the RSI study is to identify the research and development needed to build the foundation for a high-penetration renewable energy future while enhancing the operation of the electricity grid.

The RSI study consists of 15 reports that address a variety of issues related to distributed systems technology development; advanced distribution systems integration; system- level tests and demonstrations; technical and market analysis; resource assessment; and codes, standards, and regulatory implementation. The RSI reports are:

- Renewable Systems Interconnection: Executive Summary

---

* Excerpted from Subcontract Report NREL/SR-581-42304, dated February 2008.

- Distributed Photovoltaic Systems Design and Technology Requirements
- Advanced Grid Planning and Operation
- Utility Models, Analysis, and Simulation Tools
- Cyber Security Analysis
- Power System Planning: Emerging Practices Suitable for Evaluating the Impact of High-Penetration Photovoltaics
- Distribution System Voltage Performance Analysis for High-Penetration Photovoltaics
- Enhanced Reliability of Photovoltaic Systems with Energy Storage and Controls
- Transmission System Performance Analysis for High-Penetration Photovoltaics
- Solar Resource Assessment
- Test and Demonstration Program Definition
- Photovoltaics Value Analysis
- Photovoltaics Business Models
- Production Cost Modeling for High Levels of Photovoltaic Penetration
- Rooftop Photovoltaics Market Penetration Scenarios.

Addressing grid-integration issues is a necessary prerequisite for the long-term viability of the distributed renewable energy industry, in general, and the distributed PV industry, in particular. The RSI study is one step on this path. The Department of Energy is also working with stakeholders to develop a research and development plan aimed at making this vision a reality.

ACKNOWLEDGMENTS

The authors would like to thank the Department of Energy (DOE) and the National Renewable Energy Laboratory (NREL) for supporting our work on this topic and the industry experts who provided input including Robert Margolis, Peter H. Kobos, Dan Rastler, and Mark Bolinger. We would also like to thank reviewers at all of the companies who took time to review the case studies and examples presented in this document, including staff from (in alphabetical order): Arizona Public Service, Austin Energy, Borrego Solar, NSTAR, Old Country Roofing, SDG&E, the Sacramento Municipal Utility District (SMUD), SunEdison, SunPower, Xcel Energy, and We Energies.

We would also like to thank our colleagues at Navigant Consulting Inc. who provided valuable input, including Craig McDonald and Stan Blazewicz.

In addition, we would like to acknowledge that in our informal conversations with industry on the topic of future business models, we confronted a great spectrum of opinions regarding how things will unfold; in particular, this was true for utility involvement. Some industry leaders thought that utility involvement in distributed PV will remain limited, while others view utility involvement, including control and ownership, as inevitable. This chapter examines the spectrum of options for the future.

EXECUTIVE SUMMARY

As photovoltaics (PV) demonstrate the potential to significantly penetrate the electric generation market, a question arises: How might government action encourage business models that promote the development of PV?

This chapter is a first structured, comprehensive, and public attempt to answer that question. Our investigation identified several key findings:

- The question is dynamic, and has broad implications for a wide array of stakeholders—most notably utilities.
- While the number of installed distributed PV systems will eventually become a material and operational concern—or opportunity—for utilities, the full benefits of an extensive distributed PV resource are not likely to be realized without some degree of utility control and possibly ownership.
- Who owns and controls the PV facilities and the related flows of cash and other benefits is key to determining the potential viability of any PV business model.
- It appears that key industry stakeholders have considered changes to current models of ownership and control, but few have moved forward, indicating that barriers, such as the current regulatory structure, insufficient scale, and other priorities, impede optimum development.
- Smart-grid technologies are expected to be very important for the emerging PV business models. While this chapter does not focus on specific recommendations, it is clear that the ongoing RD&D in this area, both public and private, will be critical for distributed PV. Similarly, distributed PV may become an important enabler for deployment of these technologies, as higher levels of PV market penetration necessitate their use.
- Each potential future business model identified in this chapter has several permutations, and it is not yet clear which is likely to be the most successful, how multiple business models could co-exist, or if one may evolve into another over time. Attempting to pilot any particular one at this time appears to be premature.
- The scale of a potential pilot program involving utilities feeds back into the advisability of delaying the implementation of a pilot until a greater level of stakeholder engagement is achieved in the preliminary assessment.
- It appears to be a question of when, and not if, there will be a need for new PV business models, in order to accommodate and facilitate widespread adoption of distributed PV.

BACKGROUND

Current PV business models principally revolve around the ownership of PV systems by individuals and increasingly by third parties, rather than by utilities. At today's low levels of market penetration, distributed, grid-connected PV is not a central concern nor even of great interest to most utilities. However, as PV market penetration accelerates, utilities will become critical stakeholders, driven primarily by concerns about grid operation, safety, and revenue erosion.

Until now, utilities have mainly responded to regulators who asked of them nothing more than to help customers who wanted to purchase or acquire a PV system. In the process, some utilities have removed key barriers to PV deployment to a limited extent, mainly by providing net metering and adopting simplified, standardized interconnection standards and agreements. In addition, regulators have sometimes obligated utilities to purchase renewable energy certificates (RECs) generated by PV owners, particularly in states with specific mandates for solar energy.

On the whole, however, the utility's role in the PV market has been passive. PV has not been a core utility business endeavor nor a concern, primarily because 1) the cost of PV has exceeded that of other energy delivery options, and 2) utilities have seen, up to the present, no clear business/regulatory model that will allow them to recover high distributed PV costs.

Project Scope and Objectives

The objectives of the study presented here are to:

- Document current and emerging PV business models,
- Identify a range of potential future business models that enhance the value of PV to key stakeholders and thus increase market penetration (e.g., by incorporating energy storage, controls, and other technologies which allow the system to be independently controlled and dispatchable), and
- Discuss how promising potential future business models might be encouraged in the marketplace by government action, including DOE-sponsored research, development, and deployment (RD&D).

The basic premise explored in this chapter is that large amounts of distributed PV create a new paradigm that has the potential to radically alter a utility's business model. Of all stakeholders involved, it is the utility that will have its existing business model most disrupted as the PV market expands. However, it is also the utility that has the potential to best utilize the unique, quantifiable benefits of the electricity generated by a PV system.

OVERVIEW OF PV BUSINESS MODEL EVOLUTION

The PV industry is moving away from the early approach in which the customer not only owned and financed the PV system, but also managed most aspects of installation. This approach is referred to as the Zero Generation PV business model; its attractiveness was limited to a relatively small group of so-called pioneers[4] who were committed to PV's environmental, energy security, and self-generation benefits. The PV industry has evolved to 1st Generation PV business models, in which the product is more attractive to a

---

[4] Geoffrey Moore, *Crossing the Chasm*, Harper Business, 1991.

broader market, moving into the so-called early adopter customer category[5] (See Figure ES-1-1).

2nd Generation business models have yet to emerge, but will emphasize greater integration of the PV systems into the grid because emerging technologies and regulatory initiatives are likely to make such integration more viable and valuable. 2nd *Generation* business models are the focus of the future business models explored in this chapter, as they are expected to become increasingly important to various stakeholders.

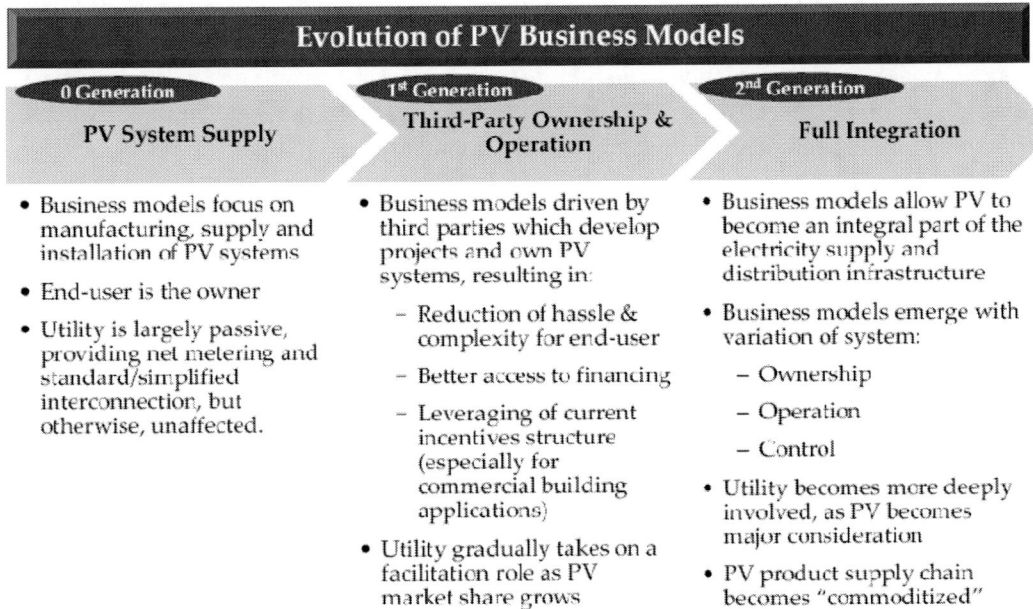

Evolution of PV Business Models

0 Generation	1st Generation	2nd Generation
PV System Supply	**Third-Party Ownership & Operation**	**Full Integration**
• Business models focus on manufacturing, supply and installation of PV systems	• Business models driven by third parties which develop projects and own PV systems, resulting in:	• Business models allow PV to become an integral part of the electricity supply and distribution infrastructure
• End-user is the owner	– Reduction of hassle & complexity for end-user	• Business models emerge with variation of system:
• Utility is largely passive, providing net metering and standard/simplified interconnection, but otherwise, unaffected.	– Better access to financing	– Ownership
	– Leveraging of current incentives structure (especially for commercial building applications)	– Operation
		– Control
	• Utility gradually takes on a facilitation role as PV market share grows	• Utility becomes more deeply involved, as PV becomes major consideration
		• PV product supply chain becomes "commoditized"

Figure ES-1-1. Evolution of PV business models.

Although the utility to date has been generally *reactive* to state requirements (e.g., net metering, standardized interconnection), it is expected to become *proactive* in the distributed PV market as it is pushed to key stakeholder status. Once PV reaches significant market penetration (perhaps 10-15% of a utility's peak load), utility involvement will be driven by concerns for grid infrastructure, safety, and of course, revenue erosion. An appropriate business model can promote and accelerate the utility's willing promulgation of PV and help unlock its full value.

Current PV Business Models

In the PV marketplace today, numerous interesting developments are occurring within the confines of the existing regulatory framework and utility business structure. The growing PV industry is developing new, more efficient ways to deliver products, services, and financing to customers while also addressing key market barriers. Improvements and

---

[5] Id.

innovations being made by the PV industry to *Zero* and *1ˢᵗ Generation* business models are very important and are a prerequisite for the industry to achieve a higher level of maturity and scale. The ongoing PV activity within these *Zero* and *1ˢᵗ Generation* business models will help bring the industry down the cost curve and up the market penetration curve, implying more availability of PV to the grid in general.

Digging deeper into how PV business models are characterized, we defined the marketplace for this study using the aspects of 1) ownership and 2) application. These two aspects were chosen for the following reasons. First, changes to system ownership have been and are expected to be a key driver of additional market growth. Second, current business models vary significantly by the application, as much of the focus of *Zero* and *1ˢᵗ Generation* business models is on the supply chain and getting product, financing and related services to customers in more effective ways. Figure ES-1-2 identifies the key current ownership-application models and their relative level of development in the market.

Leading PV Ownership-Application Models Today					
Application		Ownership			
		End-User	3ʳᵈ Party	Utility	
Residential	Retrofit	●	Minimal activity	Minimal activity	
	New Construction	○	Minimal activity		
Commercial	Retrofit	●	◉	Minimal activity	
	New Construction	Minimal activity	Minimal activity		
Grid-sited (utility side of meter)		NA	○	Minimal activity	

Emerging ○   Somewhat Established ◉   Most Established ●

Figure ES-1-2. Leading PV Ownership-Application models today.

By characterizing the current marketplace using the ownership and application framework, several things become apparent:

- The dominant ownership model has been end-user owned, *0 Generation,* models.
- The dominant applications have been both commercial and residential retrofit, not new construction or grid-sited.
- Today third-party ownership is quickly becoming an established ownership approach for commercial applications, and is also emerging as an ownership approach for grid-sited applications. This is moving the marketplace into the *1st Generation* PV business model discussed above.
- Utility ownership of distributed PV has been minimal, as viable business models have been lacking. *2ⁿᵈ Generation* models have yet to really take hold.

Each type of five key ownership-application models in Figure ES-1-2 is described graphically in the full report using a value network to show the relationships and the value transferred between key stakeholders. An example of a value network is shown in Figure ES-

1-3 for the end-user owned residential retrofit application to describe the most basic approach. Variations to the basic ownership-application models are also provided in the full report, as they provide insight into where innovation is occurring in the marketplace and can be largely correlated with industry trends, such as: reduction of hassle and complexity, product supply chain efficiency, and reduction of financing cost.

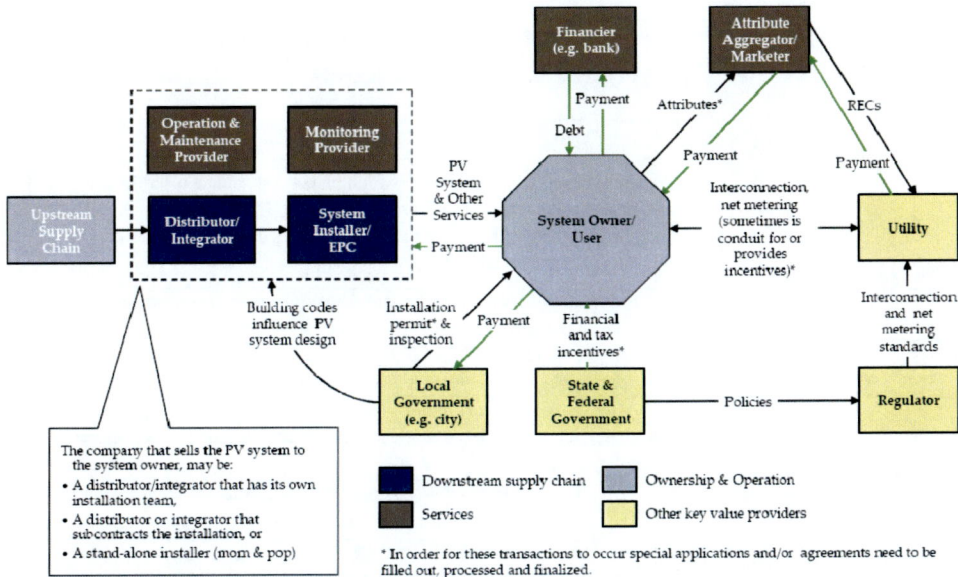

Figure ES-1-3. Example of Value Network: End-user Owned Residential Retrofit.[6]

The diagrams of the business models were useful in the study when considering how relationships and flow of values between key stakeholders would change in the future due to the emergence of new business models.

As this study progressed, it became apparent that the ownership attribute remained a key factor in defining future business models, but that the characteristic of application was not critical. Instead the question of who controls the PV system became a defining factor of future business models, thus, future business models are characterized in this chapter by ownership and control. As discussed in the future business models section below, we assume that as the PV supply chain matures the application (e.g., where the system is installed) will become less important than who controls the system and how.

Current Utility Involvement in PV Business Models

As discussed above, utilities do not currently generally own PV system. However, utilities are involved in the current PV business models in ways that are important when considering what they might do in the future. Historically, utility participation in the PV market was limited to a few cases of retailing PV systems and providing system rebates, especially by

---

[6] EPC = Engineer, Procure, Construct

municipal utilities. However, a growing number of investor-owned utilities have recently taken on more active roles in PV markets. Using our taxonomy of PV business model evolution, this type of utility activity falls under *1st Generation* utility facilitation of PV business models. This activity is seen as supporting other business models, not as a stand-alone business model. Table ES-1 below provides a brief description of current initiatives that utilities have created to facilitate the development of PV.

Table ES-1. Examples of Utility Programs Supporting Current PV Business Models

Program	Utility	Brief Description
Financing	APS	Building a structure through which banks and lenders offer special financing or refinancing to solar customers (e.g., APS pays the PV rebate directly to the lender and the incentive is used to buy down the interest rate or to re-amortize the loan).
Financing	PSE&G	Proposing to lend capital to end-users and solar developers for 40-50% of the project cost, which is repaid over 15 years with S-RECs, at a rate of 12.11%. S-RECs are valued at their floor price or current market price, whichever is higher.
Technology Partnership	NSTAR	Aligning with Evergreen Solar to lower the overall cost of solar generation by promoting standardized systems installed by pre-approved solar contractors.
REC Database	Xcel Energy	Partnering with Pioneer Solutions to develop a software application to view and track RECs for compliance and trading purposes.
Feed-in Rate	We Energies	Offering a feed-in tariff of 22.5 ¢/kWh. Eligible PV systems must be between 1.5kW and 100kW, and customers must enroll for a 10-year contract.

In addition to utility programs that support current business models, there are a few existing cases of utility ownership that could pave the way for greater involvement in distributed PV and a more active role either in ownership or control of PV systems (table ES-2). This type of activity falls into our definition of *2nd Generation* business models as they include aspects of utility ownership.

Table ES-2. Examples of Utility PV Ownership

Program	Utility	Brief Description
Customer- sited PV	SDG&E	As part of its sustainable communities program, the utility is installing PV systems at customer-sites on the utility side of the meter. Program was approved by regulators and SDG&E has $4.3 million per year. SDG&E is working with LEED certifying body so that utility's PV systems can help the building owner achieve LEED status.
Solar Shares	SMUD	SMUD is developing a grid-sited 1MW PV system which would allow ratepayers to buy "shares" in it through a surcharge on monthly electric bills. The program aims to attract homeowners or commercial customers that want solar, but cannot install it because they rent, have shading issues, or do not have access to up-front capital.

Services Agreement (with ownership option)	Austin Energy	Considering a program to lease land to project developers. The developer would build a PV or concentrating solar system and then utilize/monetize the tax benefits, perhaps benefiting from Austin's access to low cost tax-exempt debt, after which the developer may have the option to transfer the ownership to Austin Energy. Additionally, Austin Energy is considering prepayment options as well as possible lease arrangements.

Context for Future Business Models

At the same time that the PV industry is making great strides in the deployment of PV using *0* and *1st Generation* approaches, significant activities are also occurring outside of the PV industry that have clear implications for long-term PV market penetration. In particular, changes in policy, technology, and utility regulation may hold the potential to not only create opportunities to unlock additional value from PV systems, but may simultaneously create more demand for it (see Figure ES-1-4).

Technology developments underway to manage the distribution grid more effectively will have many benefits for distributed generation, including PV. In particular, the development of distribution system automation, the transition to "smart grids," and the deployment of customer- and utility-controlled demand response are all likely to help utilities and others unlock additional value from distributed PV systems.

Policy trends that create a market for renewable energy, such as renewable portfolio standards or RPS (especially those with solar set-asides) and greenhouse gas emission caps, are gaining momentum at the state and local levels, and may ultimately culminate in much higher average state targets and, eventually, a federal-level policy.

Finally, regulatory changes in some states are altering the way a utility perceives its business. Beyond net metering and interconnection issues, performance-based ratemaking (in which incentive benchmarks, rather than budgets, determine cost recovery) and revenue decoupling mechanisms (in which rates are determined as a function of service delivery rather than as a strict return on hard assets) are being implemented to encourage energy efficiency, conservation, and renewable energy. Given these types of changes, the ability of a utility to realize revenue from rates that are based in part on reconfiguring its grid and altering its customer support to integrate PV will have obvious benefits for the further increase of distributed PV. In addition, some utilities have experimented with tariff structures to encourage desired consumer behaviors and the deployment of new technologies. For example, variations of time-of-use pricing can be very beneficial to PV economics. Also, adoption of transmission congestion pricing should have a beneficial impact on distributed PV, as the market value of distributed generation will be made plain by the congestion prices. These regulatory actions are increasingly being driven by the desire to encourage conservation or greenhouse gas reductions.

Technology	Policy	Regulatory
• Development and deployment of distribution automation technologies • Transition to "smart grids" • Continued development and deployment of other distributed generation technologies • Development and deployment of plug-in hybrid vehicles (implications for grid operations, load growth and battery technology development)	• Further development of Renewable portfolio standards (increasingly with solar set asides) • Greenhouse gas emission cap & trade programs and other climate change initiatives • State-level economic development initiatives • Growth of state solar energy initiatives and system benefits charge funds	• Performance-based ratemaking • Revenue decoupling to encourage energy efficiency and conservation • Tariff structures optimized for PV and other distributed generation • Demand response programs (customer and utility controlled)

Figure ES-1-4. External factors with implications for PV market development.

Looking forward 10 to 20 years, there is a strong case to be made that PV in distributed applications, primarily customer-sited, will become an inevitable and significant component of the electricity sector, and especially if forecasted PV cost reductions materialize. In the long-term vision presented in this chapter, PV will pass a "tipping point" beyond which it is competitive with retail power supplied by the grid. The point of wide-scale competitiveness with grid power may come sooner as a result of specific breakthroughs in technology, or later as a result of the steady march down the cost curve. In either case, this vision depends on the PV supply chain being able to ramp up capacity to meet market demand.

When PV achieves a high degree of market penetration, there will be significant implications for key stakeholders, especially for the utility. PV will eventually be an operational problem for the utility if it is not strategically managed. Additionally, as the cost of PV comes down, distributed PV generation could become a competitive threat to central-station generation.

Of all the stakeholders involved, it is the utility that will have its existing business model disrupted the most, and must therefore adapt to protect and enhance its business. Thus, greater utility involvement is seen as the key to future PV business models. In contemplating PV system ownership and control on the distribution grid, a utility can leverage what it already does well, including asset management and investment, customer service, and system operations (see table ES-3).

For the utility, PV could simply become another rate-based asset to own, manage, and operate to provide equal or higher quality of service than what it provides today. In addition, PV may allow a utility to take maximum advantage of the capabilities that distribution automation and smart grid technologies will provide. In fact, PV arguably has the potential to be one of the most significant distributed resources managed by these technologies.

**Table ES-3. Implications of Widespread Distributed
PV Deployment on Key Stakeholders**

Stakeholder	Implications
End-user	PV system: • Cost-effective alternative to the grid • Provides improved reliability (over grid) • Helps meet environmental desires of consumers • Generates a range of value streams (driven in part by environmental and climate change policy) • Part of a bundle of new technologies to improve energy service at end use and reduce cost as cost drops (low-cost energy storage, distribution system automation, "smart homes", plug-in hybrid vehicles)
System Owner	• PV system output has multiple value streams that can make it competitive in the market relative to grid power • Owner needs to be able to identify and capture multiple PV value streams

Distribution Utility and Vertically Integrated Utility	• High degree of PV market penetration creates:
	
	  o Reduced throughput leading to revenue loss under traditional tariff structures
	  o Need for control of PV systems and/or new distribution system architectures to ensure safety, operational integrity, and reliability of the distribution grid
	
	• In addition, new technologies used in conjunction with PV could radically change utility operations and product/service offerings to

Wholesale Generator	• High degree of PV market penetration could provide competition in the wholesale market to more expensive generating assets
Regulator	• Emergence of cost-effective PV and other complementary technologies creates need for major transformation of how utility industry is regulated
Transmission Company	• High degree of PV market penetration could impact the demand for transmission services

FUTURE BUSINESS MODELS

Three basic types of business models were identified in this chapter, as illustrated in Figure ES-1-5. The main distinctions among them lie in who owns and controls the PV system (a fourth option, in which the PV system is owned by the utility but not controlled by it, is not viewed as being a viable business model because the utility is unlikely to cede control of an asset that it owns). As will be discussed in more detail below, the success of any of these three business models will be tightly linked to ongoing technology and market developments in distribution automation and demand response, and may also require significant regulatory changes. In the full report, variations of each basic type of business models are discussed.

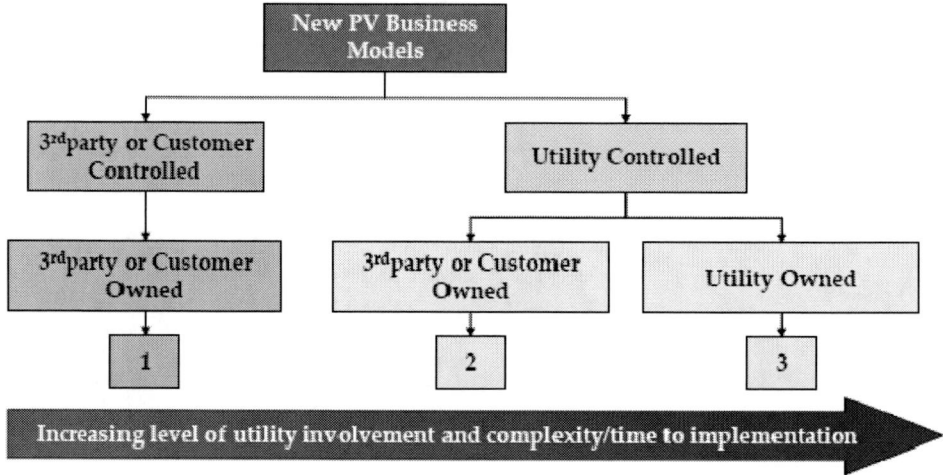

Figure ES-1-5. New PV business models focused on system ownership and control.

Third Party/Customer Controlled and Owned PV Business Model

In this business model, the customer or a third party controls the PV system as well as owns it (there is also the possibility of customer ownership combined with third-party control). This business model is primarily an extrapolation of current business models and trends (Figure ES-1-6). The key difference is that additional sources of revenue are captured by the owner, based on various changes to the regulatory and policy regimes, and on the deployment of "smart grid" technologies and energy storage that is integrated with PV system operation. In this model, the utility role remains mainly one of facilitation, primarily driven by regulatory or policy changes. The utility pays for value-added products and services obtained from the PV system and are then allowed to recover these costs through traditional rate-making proceedings.

This business model is considered the most likely to become established in the absence of outside influence, as various pieces of current regulation and policy are already in place to enable it in some jurisdictions.

If the customer/third-party controlled and owned business model becomes widespread, the distribution grid must be re-engineered to be highly responsive to changes in PV operating profiles (e.g., extremely localized power fluctuations), either due to transient changes in sunlight availability or to decisions taken by the owners, because the utility will not control the PV systems. An issue that will arise is the degree to which owners will be "free to choose" how to operate their systems. For example, if a customer chooses to participate in a demand response program, they might be obligated to respond to utility signals.

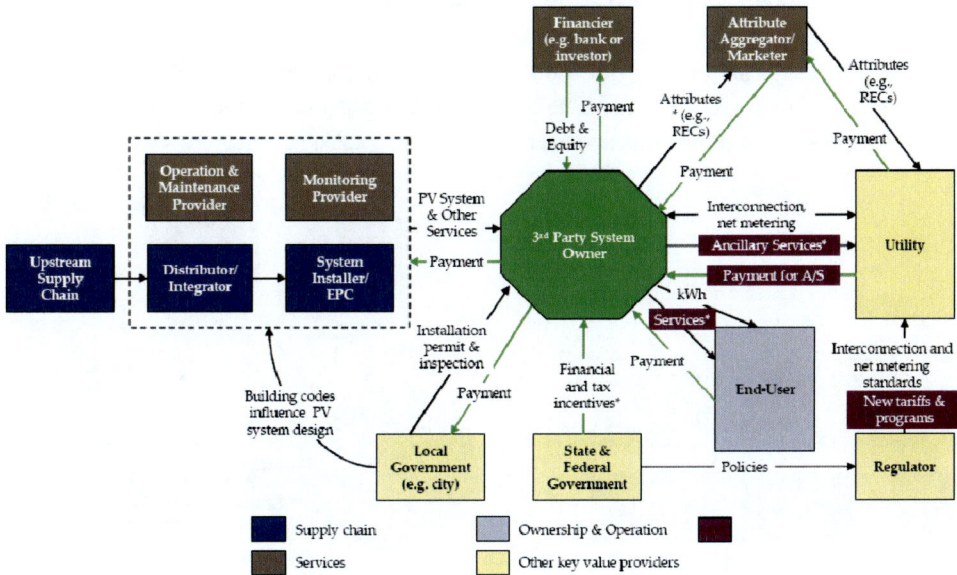

* Requires communications and control, including for performance - based incentives.

Figure ES-1-6. Third-party/Customer Controlled and Owned Value Network.[7]

Utility Controlled, but Third-party or Customer Owned PV Business Model

This business model is somewhat similar to the one described above, in that it seeks to achieve similar objectives (Figure ES-1-7). The key difference is that greater utility involvement in the operation and control of the systems is thought to be a way to increase the value of the assets. Like the customer controlled business model described above, regulatory and policy regimes will need to change, though more significantly here, to allow the utility to reach behind the meter where the PV system will reside. In this case, the customer will not respond to price signals because the utility is controlling the PV system, at least to some extent.

This business model may work best where aggressive demand response or other similar programs are being pursued, or where high penetration of PV systems may pose serious grid control and operations issues. Under those circumstances, direct utility control—for example, to allow the utility to curtail PV system operation to maintain grid stability— instead of a complicated market for such services, may be preferable because the utility is assured response as it controls the asset as opposed to relying on optional responses to price signals.

In this model, the utility would still pay for value-added products and services from PV systems and then be allowed to recover these costs through traditional rate-making proceedings. To the extent that PV systems provide a service and create value (e.g., avoid costs) for the utility, this would be factored into the cost of recovery calculation.

---

[7] As an example of this business model, we selected the third-party owned variant, although the end-use owned variant would illustrate the issues as well. In addition, this diagram represents all of the major functions as separate, even though there may integration of some functions as the industry grows and matures.

This business model is expected to evolve more slowly given the additional regulatory changes required to permit utility control behind the meter. Additionally, distributed PV needs to exist at a significant scale in order for a utility to find value in controlling it. For example, the distributed PV installation would have value to the utility proportional to its capacity to substitute for generation, capacity, and transmission and distribution (T&D) investments.

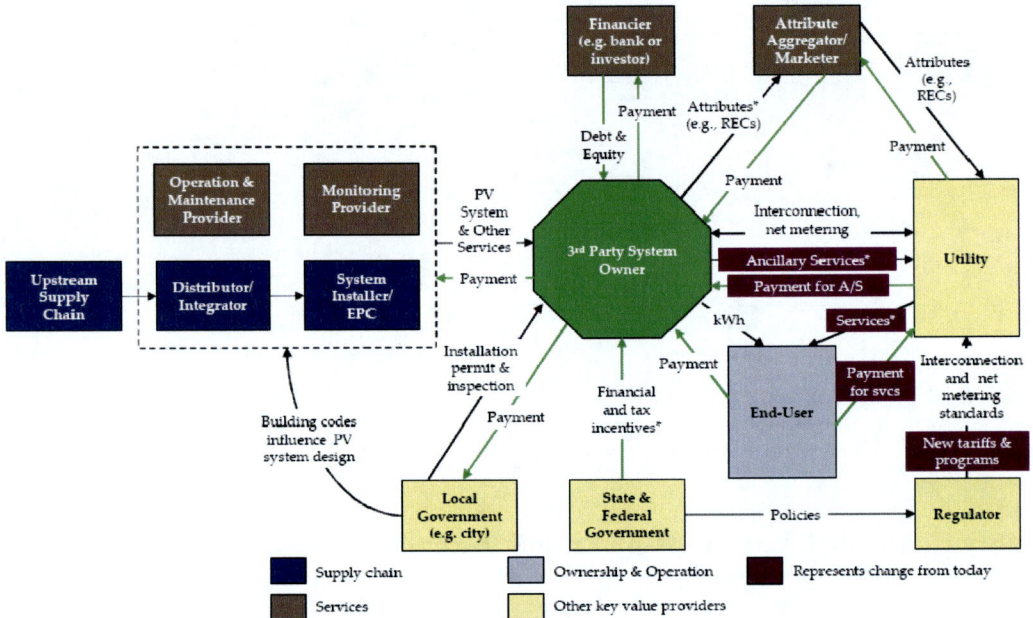

* Requires communications and control, including for performance - based incentives.

Figure ES-1-7. Utility Controlled but Third-party/Customer Owned Value Network.[8]

The requirements for the utility controlled, customer/third-party owned business model are largely the same as for customer/third-party controlled model. The key difference is the regulatory regime, which would enable utility to control significant assets on the customer side of meter. To the extent that utility control is not just for grid benefits but also to enable the utility to offer other services to the end user, these regulatory changes will need to address the rules governing competition for providing these services. The main competitive issue is that the utility, as a monopoly, has an unfair advantage in its access to the customer. If the utility is allowed to access assets behind the meter for the benefit of the grid, but then is also allowed to leverage this access to offer customer- based services like backup power or energy management, other companies without such access might see this as unfair. To the extent that utilities are allowed to use the PV assets to provide value-added services to those customers who own them, the structure and pricing of these services must be determined in a transparent and equitable manner.

---

[8] As an example of this business model, we selected the third-party owned variant, although the end-use owned variant would illustrate the issues as well. In addition, this diagram represents all of the major functions as separate, even though there may integration of some functions as the industry grows and matures.

Utility Controlled and Owned PV Business Model

This business model represents the greatest departure from today, as the utility reaches unequivocally behind the meter to own assets and provide a range of services to customers (Figure ES-1-8). This model seeks to unlock greater distributed PV value by involving the utility directly in both ownership and control of the asset, and in monetization of the asset's value. This arrangement fits well with utility core competencies of asset ownership and operation. Given that PV is a capital-intensive asset, there is merit in putting such utility-owned assets in the rate-base.

By allowing the utility the greatest control over the placement and subsequent operation of the asset, this model should generate the greatest overall value for the utility. Moreover, in this model, the utility can readily incorporate the grid benefits into its basic cost of service, as well as sell value-added services to the end-user. Of the three groups of business models, this one is the easiest model for the utility to incorporate deployment of PV into their capital planning, as the ultimate decision to install is in their control. However, the issue of competition will be a complication as the utility could have unfair advantage in providing value-added customer-oriented (vs. grid oriented) services that a third party may want to provide.

Like the other business models described above, regulatory and policy regimes will need to be changed significantly to allow the utility to reach so overtly behind the meter. To mitigate the potential scope of such regulatory and policy changes, the PV systems could be located on customer premises but placed on the utility side of the meter. In the past, states have prohibited utilities from owning and operating distributed energy resources (DER) because of concerns regarding market power. This concern will need to be addressed if and when PV systems become very inexpensive or otherwise attractive to utilities.

This business model is expected to evolve more slowly than the others, given the additional regulatory changes required to permit utility control and ownership. Additionally, in order for utility control to have significant value to the utility, distributed PV has to exist on a sufficient scale to have material impact on key values such as ability to offset generation, capacity, and T&D investments.

In a business model where the PV assets are both controlled and owned by the utility, the structure of the system-wide control architecture would be different than in models where the customer or a third party either controls or owns the assets. There would be no need to be able to send signals to a large number of owners. Instead, the control of the PV assets would be integrated into the utility's overall distribution network. Moreover, the deployment and use of PV systems would be more readily integrated into the utility's planning processes; PV systems would become extensions of the distribution grid. Thus, as PV is continually added, the utility would have the opportunity to make sure that the grid configuration remains optimal. Also, this business model would likely make it easier for utilities to justify investments required for grid reconfiguration, as this becomes necessary.

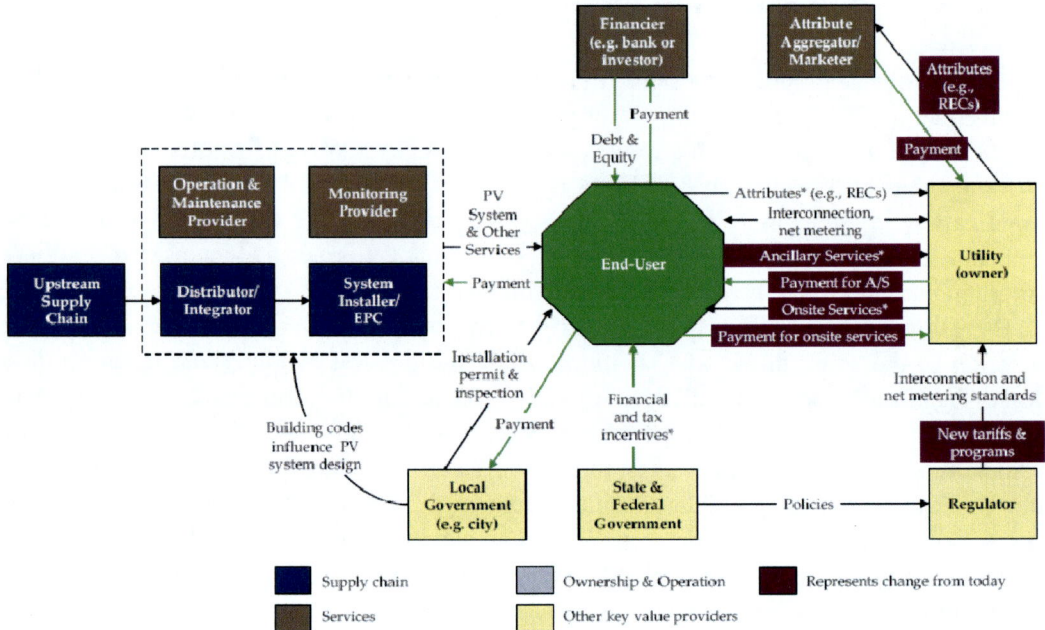

* Requires communications and control, including for performance - based incentives.

Figure ES-1-8. Utility controlled and owned value network.[9]

CONCLUSION

Currently, PV business models revolve around access to lower-cost financing, increasing the efficiency of the supply chain, and reducing hassles and complexity for the customer. These types of incremental improvements will occur naturally as *0* and *1ˢᵗ Generation* business models continue to evolve.

Up until this point, there has been little reason to address system control or consider PV aggregation as an explicit policy matter, given the limited number of PV systems installed on the distribution grid. However, a time will come—in some areas of the country, much sooner than others—when the sheer number of installed distributed PV systems becomes a material and operational concern—or opportunity—for utilities. Policy and regulatory considerations will then be paramount.

The most significant finding in this study to date is that the full benefits of an extensive distributed PV resource are not likely to be realized without some degree of utility control and ownership. The need to have active management and control of an increasingly large number of distributed PV systems implies that utilities will most likely become more involved in one way or another. As market penetration increases, distributed generation will reach a scale (i.e., generally greater than 100 MW) that could translate to significant value. For example, utility involvement could help optimize distributed PV assets by incorporating them into grid and generation planning. This is likely to reduce new peaking power requirements,

---

[9] This diagram represents all of the major functions as separate, even though there may integration of some functions as the industry grows and matures.

distribution substation upgrades, and other system investments, thus unlocking latent value in the electric grid as a whole.

The results of the analyses performed in this series of DOE studies show that the real value of PV lies in its potential to offset generation, capacity, and T&D investment. Such value greatly outweighs the value PV has for providing ancillary services on the distribution grid. Therefore, business model development will not be driven by the potential for ancillary grid services. It is the possibility that a large quantity of distributed PV systems will be installed that provides the greatest potential benefit to the nation's energy infrastructure, as these systems in aggregate could actually offset significant investment requirements in new generation, transmission, and distribution capacity.

Through its efforts on Renewable Systems Interconnection, DOE is investing in understanding how technologies on the distribution grid can make significant contributions to meeting future electricity demands. Continued work on business models is a natural complement to this, as business models will facilitate how all of the technologies ultimately come together and transfer value to stakeholders. The future business models described in this chapter will require changes to industry structure, which implies risk for key stakeholders. DOE is in a position to work with key stakeholders to help mitigate some of the risk involved in pursing these new approaches.

To understand the potential real costs and benefits, promising future business models will need to be piloted at a sufficient scale, requiring significant time and investment. Today, the exact scope, duration, and scale of the business model pilots required is not clear because many issues still need to be addressed. It is critical that key stakeholders are engaged in understanding exactly what is holding back the development of these business models, because many of these companies and organizations are actively considering, right now, what the future will look like and how they will participate. In addition, explicit business model development should be coordinated with work on smart grid capabilities (e.g., distribution automation and advanced metering infrastructure), energy load management (e.g., demand response) and other distributed resource technologies. All of the business models discussed in this chapter will require integration with these other emerging technologies and capabilities, as well as PV. Because of the potential fundamental changes in regulation, technology, ownership, control, and grid management implied by the former, it is premature at this point for DOE to issue an RFP to pilot new business models. As an alternative to immediate pilot activity, we recommend the following three-phase approach which is illustrated in Figure ES-1-9:

- Phase 1 – Build the Foundation
- Phase 2 – Develop the Scope
- Phase 3 – Pilot Business Models and Fund Other Supporting Activities

Phase 1: Build the Foundation	Phase 2: Develop Scope	Phase 3: Pilots & Other Activities
• Stakeholder engagement • Collaboration with the smart grid, energy management and non-PV DG communities • Analytical studies	• Take results from Phase 1 to develop the scope for pilot activities and identify other targeted activities to help promote promising business models	• Pilot business models • Fund other activities to development promising business models (e.g. sample regulations, analysis tools)

Phase 1

Phase 2

Phase 3

2 years   2+ years

DOE efforts on Renewable System Interconnection will inform business model pilots

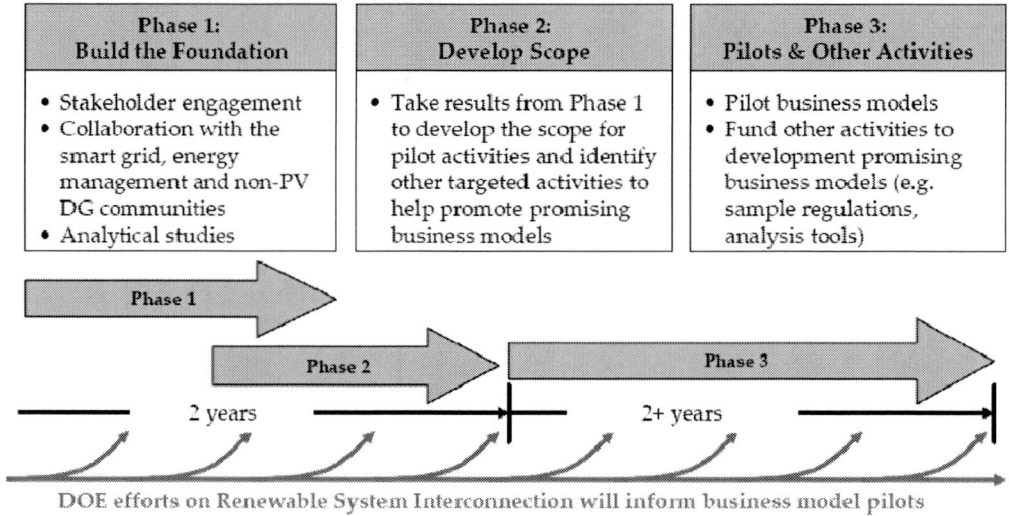

Figure ES-1-9. Three phase approach to developing business model pilots.

Work in Phases 1 and 2 involve studies and preparation for future business model pilots . These first two phases will likely take one and a half to two years. This additional time will allow the industry to mature and the pace of PV deployment to increase so that it is more likely to achieve a scale sufficient to support business model pilot s in Phase 3. These recommendations are detailed in the full chapter, which follows.

1.0. INTRODUCTION

1.1. Background and Context

U.S. grid-tied PV markets are growing rapidly. Between 2001 and 2006, the U.S. grid- tied PV market had a 46% compound average rate of growth (CAGR). The US grid-tied market is expected to continue strong growth between 2006 and 2011 with a CAGR of 50%.[10] Nevertheless, in relation to the overall size of the power sector, the PV market is very small: <<1% of total installed capacity. As such, the effects of distributed PV on the electricity grid today are minimal. This applies to both ongoing utility operations and utility planning (capacity, transmission & distribution investments). Consequently, utilities are generally not involved in the PV value network, except as it relates to the provision of net metering tariffs and standardized/simplified interconnection, and even then, these services are generally offered to fulfill to a regulatory requirement, with the utility maintaining a largely passive role once the PV system is in operation.

However, as PV market penetration grows, its presence on the grid will begin to have a larger impact on grid operations and planning, and more generally, on how utilities conduct their business. This applies to vertically integrated utilities and distribution

---

[10] NCI PV Services, September 2007.

companies in unbundled power markets, as well as other key players in the electric power sector, including wholesale generators, power marketers and transmission companies.

As the PV value network matures and becomes more efficient, there are policy, technology, and regulatory changes that will enable and/or require higher market penetration of distributed PV. This suggests that the business models that characterize the current PV value network will continue to evolve and that new business models may also emerge, particularly those that have greater levels of utility involvement. Ideally, these new business models should do more than simply respond to potential problems that arise from higher levels of PV market penetration, but anticipate them and thus mitigate or avoid them altogether.

Today, leading edge PV business models focus on third-party ownership, primarily as a means of addressing the barrier of the high up-front cost of the systems. This improves access to greater amounts of lower-cost capital, optimizes the value of existing incentives and therefore makes the technology accessible to a broader market. These innovations, along with improvements to the end-user ownership model, are a sign of increasing sophistication and scale within the PV value network, but are largely about achieving higher levels of efficiency under current cost and incentives structures and regulatory regimes. While extremely important, these business models do not address the fundamental issues that will eventually arise as PV market penetration reaches levels where it will begin to materially impact utilities and grid operations.

1.2. Study Objectives and Scope

The objectives of the study presented here are to:

- Document current and emerging PV business models;
- Identify a range of potential future business models that enhance the value of PV to key stakeholders and thus increase market penetration (e.g., by incorporating energy storage, controls, and other technologies which allow the system to be independently controlled and dispatchable); and
- Discuss how promising potential future business models might be encouraged in the marketplace by government action, including DOE-sponsored research, development, and deployment (RD&D).

The basic premise explored in this chapter is that large amounts of distributed PV create a new paradigm that has the potential to radically alter a utility's business model. Of all stakeholders involved, it is the utility that will have its existing business model most disrupted as the PV market expands. However, it is also the utility that has the potential to best utilize the unique, quantifiable benefits of the electricity generated by a PV system.

1.3. Chapter Contents

Section 2 provides a brief review of prior work in this area. In Section 3, the project approach is discussed. Section 4 contains the results of the work, as follows:

- Section 4.1 provides background on the PV value network and defines the term "business model" as used in this chapter.
- Section 4.2 categorizes dominant PV business models today based on system ownership and application, describes current and emerging PV business models, and provides detailed case studies on current leading-edge business models and utility involvement.
- Section 4.3 articulates a future vision for high penetration of PV, with emphasis on implications for distributed generation (like PV) and utilities. It identifies potential future PV business models, considering: alternatives for ownership and control, opportunities for expanded utility roles, and use of energy storage, controls, and other advanced technology packages. It also identifies challenges for these new business models.
- Section 5 summarizes the conclusions of this study and identifies and prioritizes possible actions for government to encourage promising business models.

2.0. CURRENT STATUS OF EXISTING RESEARCH

Private industry is constantly engaged in advancing PV business models, like the ones discussed in Section 4.2 of this chapter. However, much less work has been done by government to support development and implementation of new PV business models. Recent business model research supported by the public sector includes:

- The Public Interest Energy Research (PIER) group in the California Energy Commission held a workshop in 2004 with key stakeholders and then provided grants to advance PV business models to support the Zero Energy New Homes Initiatives.
- The Electric Power Research Institute (EPRI) conducted a program in 2006 and 2007 called STAC, to create and demonstrate utility incentives for distributed energy resources, including PV. The work was supported by California Energy Commission, Massachusetts Technology Collaborative, and the State Technologies Advancement Collaborative. The stakeholders involved in the project included government, utilities, vendors, developers, consumer representatives, and public interest groups. No pilots as of yet have resulted from this project and the results of the study are not yet available to the public.
- The Department of Energy (DOE) is currently supporting the Solar Electric Power Association (SEPA) to develop new PV utility business model concepts via working groups. The results from the working group are expected in early 2008.

3.0. PROJECT APPROACH

Using in-house experience and drawing upon prior work in the area of PV business models, the starting point for the analysis was to identify and describe dominant and emerging PV business models. This work also entailed describing the PV product supply chain and the value networks[11] associated with the various business models. We also identified companies that would be profiled to provide case studies for the key business models.

We reviewed several recent reports and presentations on both PV business models and distributed generation, more generally. Our initial characterizations were also reviewed with various stakeholders via telephone interviews and in person at the Solar Power 2007 conference.

Each of the main business models, dominant and emerging today, were then characterized in a structured way to help identify barriers, important business model variations, and trends.

The results of these characterizations then served as the basis for developing hypotheses of what new business models would look like. An important part of developing future business models was to articulate a long-term vision for PV, with an emphasis on the implications for electric utilities faced with high levels of PV deployment on the distribution grid (i.e., >10% of peak demand).

Using internal expertise and subject-matter experts in PV, distribution system automation, energy management, and utility trends, we developed a structured view of possible future business models, emphasizing alternatives for ownership and control, opportunities for expanded utility roles, and use of energy storage, controls, and other advanced distribution system technologies. These future business models were also reviewed with external stakeholders.

Finally, based on the detailed characterizations of these business models, we identified the challenges for implementation and developed recommendations for research, development, demonstration, and focused on ways that government action might encourage the promising business models.

4.0. PROJECT RESULTS

4.1. Definition of PV Business Models and Industry Structure

4.1.1. PV Product Supply Chain and Value Network

Before current PV business models could be defined and analyzed, it was necessary to identify on which part of the PV value network this chapter would actually focus. This was done by first mapping the upstream and downstream components of the PV product supply chain, from the raw silicon material input to the final system installation output. Although growth and investment in the upstream supply chain are currently a major focus of the industry, this chapter focuses on the downstream portion of the supply chain—any activity following the module assembly and first point of sale into the marketplace. This is where grid impact is determined and where different models of system operation and control will occur, both issues of interest to DOE (Figure 4-1).

---

[11] The value network describes the relationships between the different stakeholders, and describes how business creates, sells, and delivers value to customers.

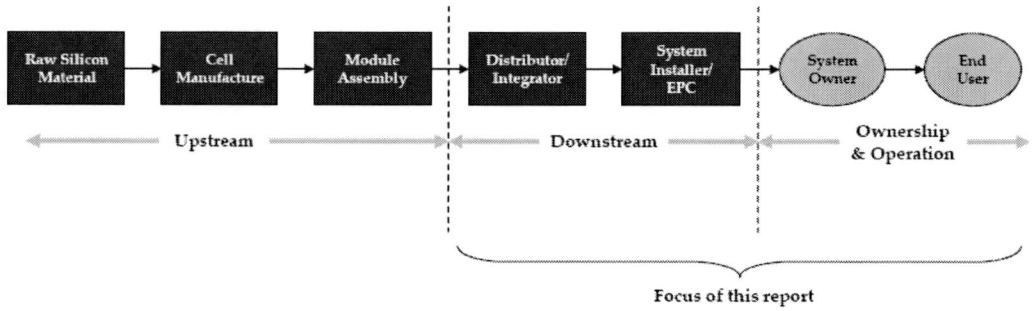

Figure 4-1. Basic components of PV product supply chain.

Considering more closely the downstream portion of the supply chain, there are several additional services and participants that play important roles in the PV marketplace. Figure 4-2 below illustrates the expanded value network beyond the PV product supply chain. The value network incorporates key services and participants that add value or otherwise exert influence on the development of the PV market and the types of business models that can exist. The red box indicates the focus of this chapter, which is primarily around models of PV ownership and operation, which include the end-user.

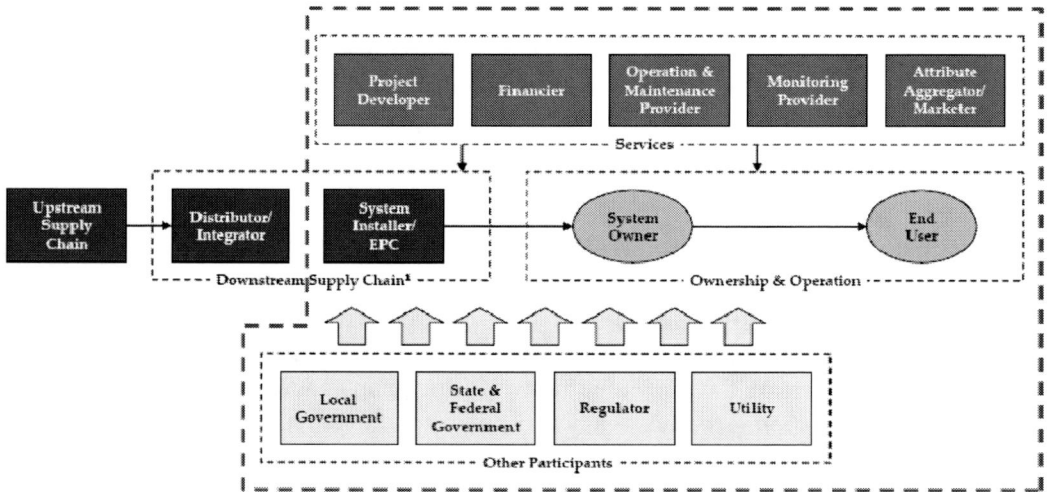

Figure 4-2. PV product supply chain and value network.

In the value network in Figure 4-2:

- For clarity, each element of the *value network* is shown separately, but many could be combined. In particular, the System Owner and End User can be the same entity or separate.
- The System Installer/EPC (Engineer, Procure, Construct) is included insofar as they may take on an ownership/operations role or provide other services.
- Similarly, a Downstream Supply Chain player can simply provide the PV system and installation, or also provide Services and/or interact directly with or be

engaged in some way with the Other Participants (e.g., by applying for building permits from Local Government and incentives from State Government).

- Operation, Maintenance, and Monitoring are value-added activities that can be provided by multiple players in the value network, but may not be viable as standalone businesses due to the limited revenue potential. Unlike other power generation businesses, operations and maintenance of PV systems are not two very distinct activities.
- The Attribute Marketer provides a monetary value for the unique attributes of the electricity generated by the PV system (e.g., solar renewable energy credits [SRECs], emissions credits). In the future, this could also extend to generation, transmission, and distribution cost avoidance benefits and grid operations benefits such as voltage support and peak shaving. Various participants in the value network could purchase these attributes.

4.1.2. Business Models Defined

A business model, for the purpose of this chapter, is defined simply as "how a company makes money." This is an important distinction because certain stakeholder activity, a state rebate for example, facilitates the growth of the PV industry, but is not technically a business model. Similarly, situations whereby states may create a market for S-RECs or mandate the inclusion of solar on new homes are not new business models. The profit- earning business activities that develop around these programs, whatever they may be, are business models.

4.1.3. Evolution of PV Business Models

Despite many years of proven field experience and recent growth, PV, as an industry, is still in its relatively early development stages, and still has a significant way to go until it reaches the scale of other similar industries. Globally, the wind power industry is larger, in MW terms, by about an order of magnitude. Other industries that can be considered comparable from a building systems perspective (e.g., HVAC) are much larger and mature, from a product and supply chain perspective, as well as from an installation and maintenance perspective.

The PV industry is moving away from the early approach in which the customer not only owned and financed the PV system, but also managed most aspects of installation. This approach is referred to as the Zero Generation PV business model; its attractiveness was limited to a relatively small group of so-called pioneers who were committed to PV's environmental, energy security, and self-generation benefits. The PV industry has evolved to *1st Generation* PV business models where the product is more attractive to a broader market, moving into the so-called early adopter customer category (see Figure 4-3).

2nd Generation business models have yet to emerge, but will emphasize greater integration of the PV systems into the grid because emerging technologies and regulatory initiatives are likely to make such integration more viable and valuable. *2nd Generation* business models are the focus of the future business models explored in this chapter, as they are expected to become increasingly important to various stakeholders.

Although the utility to date has been generally *reactive* to state requirements (e.g., net metering, standardized interconnection), it is expected to become *proactive* in the distributed PV market as it is pushed to key stakeholder status. Once PV reaches significant market penetration (perhaps 10-15% of a utility's peak load), utility involvement will be driven by concerns for grid infrastructure, safety, and of course, revenue erosion. An appropriate business model can promote and accelerate the utility's willing promulgation of PV and help unlock its full value.

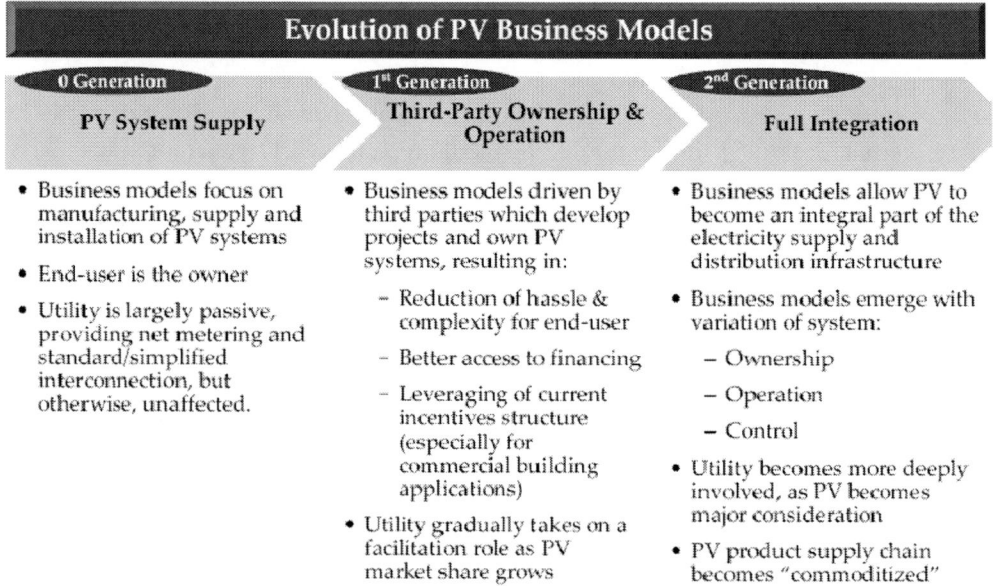

Figure 4-3. Evolution of PV business models.

4.2. Current PV Business Models

4.2.1. Overview of Approach

The analysis of current PV business models began by identifying dominant and emerging models based on the type of system ownership and the application. These two aspects were chosen for the following reasons. First, changes to system ownership have been and are expected to be a key driver of additional market growth. Second, current business models vary significantly by the application, as much of the focus of *Zero* and *1st Generation* business models is on the supply chain and getting product, financing, and related services to customers in more effective ways.

Using this approach, dominant and emerging business models are described and variations and trends for each model are identified. Finally, utility programs and state initiatives that are facilitating the development of the PV market are described. While these utility and state activities are not considered business models, they are important in understanding current utility involvement and increased emphasis by states to promote PV today.

4.2.2. Types of Ownership and Application

4.2.2.1. Ownership
The three types of ownership considered in this analysis are: system-user, third-party, and utility. A brief description of each is provided in table 4-1.

Table 4-1. Possible PV System Owners

Owner	Description
System-user	• Generally the owner of the building where the PV system is installed (e.g., a home or commercial facility) and/or the main user of the power from the PV system.
	• Traditionally, the system-user has been the dominant owner-type of PV systems in the U.S. System-users were often the pioneer and early-adopter customers that were motivated to purchase a PV system based on its noneconomic attributes (e.g., clean energy, independence from utility).
Third-party	• Not the system-user or the utility, this is another party that owns the PV system and then sells the power or use of the system back to the owner or user of the building where the PV system is installed.
	• This is emerging as a powerful owner-type, since a third-party often has access to low cost financing, greater ability to take on/understand/mitigate technical risks, and ability to make use of all government incentives.
Utility	• Utilities are central to this analysis as they are critical stakeholders to involve in the PV market as it grows;
	• However, utility ownership of distributed PV has been minimal.
	• Part of the barrier to utility ownership has been the state and federal incentives, for which utilities are not generally eligible (e.g., 30% Federal Investment Tax Credit, state rebates and performance incentives in California).
	• Utilities have not, up to now, perceived great value in distributed PV.

4.2.2.2. Application
In the United States, there are two major applications for grid-tied PV: residential and commercial. Utility owned applications (on the utility side of the meter) exist but are still extremely limited in comparison. Of the nearly 2,000 MW of PV deployed worldwide in 2006, grid-tied residential and commercial comprised 42% and 43%, respectively.[12]

Residential and commercial applications can be further segmented into markets for both new construction and existing (retrofit) buildings. Descriptions and typical system sizes for each type of market are provided below in Figure 4-4.

The residential retrofit market is currently the largest, in terms of installed capacity; however, it is one of the slowest growing today relative to other applications. The strongest growth, at greater than 40% per year, has been seen in the commercial retrofit market. This can be attributed to the emergence of third-party ownership and more attractive economics for system-users (due largely to federal incentives). The remaining three markets—residential and commercial new construction and grid-sited applications—are still relatively small but showing steady growth.

---

[12] Navigant Consulting PV Service Program, August 2007.

Key Grid-Connected Applications in the US			
Application		Description	Typical Size Range
Residential	Retrofit	• PV modules installed on roofs of existing homes; average system size increasing	2-8 kW
	New Construction	• Increasingly, building integrated products are installed at time of home construction	1.5 – 5 kW
Commercial	Retrofit	• PV modules installed on roofs of existing commercial and industrial spaces; average system size increasing	20 – 1,000 kW
	New Construction	• Increasingly, building integrated products installed at time of building construction (especially in Europe)	20 – 1,000 kW
Grid-sited		• Ground-mounted PV systems installed on the wholesale side of the distribution grid	> 500 kW

Figure 4-4. Key grid-connected applications in the United States.

There are five key ownership-application models, as depicted in Figure 4-5. By characterizing the current marketplace using the ownership and application framework, several things become apparent:

- The dominant ownership model has been end-user owned, *0 Generation,* models.
- The dominant applications have been both commercial and residential retrofit, not new construction or grid-sited.
- Today, third-party ownership is quickly becoming an established ownership approach for commercial applications, and is also emerging as an ownership approach for grid-sited applications. This is moving the marketplace into the *1st Generation* PV business models discussed above.
- Utility ownership of distributed PV has been minimal, as viable business models have been lacking. *2nd Generation* models have yet to really take hold.

Leading PV Ownership-Application Models Today				
Application		Ownership		
		End-User	3rd Party	Utility
Residential	Retrofit	●	Minimal activity	Minimal activity
	New Construction	○	Minimal activity	
Commercial	Retrofit	●	◉	Minimal activity
	New Construction	Minimal activity	Minimal activity	
Grid-sited (utility side of meter)		NA	○	Minimal activity

Emerging ○   Somewhat Established ◉   Most Established ●

Figure 4-5. Leading PV Ownership-Application models.

4.2.3. Comparison of Current Business Models

This section compares four of the leading models shown in the Figure 4-5 including:

- End-user Owner/Residential Retrofit
- End-user Owner/Residential New Construction
- Third-party Owner/Commercial Retrofit
- Third-party Owner/Grid-sited

These models are compared across value networks, variations, trends, and key characteristics, such as: customer profiles, system characteristics, marketing processes, and sources of financing. Value networks were used to show the relationships and value transfer between key stakeholders. The graphic depictions of the value networks were also useful in the study when considering how relationships and flow of values between key stakeholders may change in the future due to the emergence of new business models. Variations on each basic model are included since they provided insight into where innovation is occurring in the marketplace and can be largely correlated with industry trends. In addition to the generic characterizations, the Appendix contains case studies for the four business model types highlighted in this section.

Although the End-user Owned/Commercial Retrofit model is well established, it is not included in the comparison in the next section since it is somewhat similar to the End- user Owned/Residential Retrofit model and Third-party Owned/Commercial Retrofit model that will be described.

4.2.3.1. Comparison of Key Characteristics

Table 4-2 compares key characteristics of the four business model types highlighted in this section.

4.2.3.2. Comparison of Value Networks

The value network for the End-user Owned/Residential Retrofit model is shown in Figure 4-6. The diagram shows the flow of values and key transactions between the main stakeholders in its most basic configuration. Variations of this basic model are found in the marketplace, and typically emerge from players merging together to offer services (e.g., financing flowing through the system provider), or a stakeholder taking on responsibility for specific transactions (e.g., system provider managing interconnection with the utility, rebates with the state government, and permits from the city).

Table 4-2. Leading Business Models

	Leading Business Models			
	End-user Owner: Residential Retrofit	End-user Owner: Residential New Construction	Third-party Owner: Commercial Retrofit	Third-party Owner: Grid-sited
Customer Profile	• Customers are typically early adopters, motivated by attraction to technology, energy independence, and environmental benefits (despite all of this, incentives are still critical to the market)	• Customer of PV is the buyer of a new home; PV is sometimes seen as a favorable, additional asset, especially where rebates and environmental sentiment are strong (e.g., California) • In some cases, PV comes as a standard feature, giving the homebuyer no choice in the decision	• Motivation of customer purchase can be complicated due to the number of parties potentially involved; motivations generally include potential electricity cost savings or hedge against future electricity rate escalation and recognition for environmental leadership (e.g., brand) • Customer may still have aspects of early adopters' behavior, motivated by attraction to technology, energy independence, and environmental benefits, but may also be identified as early majority (since economics with incentives can be competitive and most risk is taken by third party). • Demand would be significantly lower without subsidies and incentives	Grid-sited • Utility is customer for power; driven by need to comply with RPS policies • Given PV's levelized cost of electricity versus distributed power, this model currently exists purely for compliance purposes
System Characteristics	• Average system 4.5-kWp and even larger in parts of U.S. • Typically covers 80% of end-user annual home load • Primarily crystalline silicone modules (~150-300W) and three-phase inverters • Use of storage and controls has been minimal (although prevalent in off-grid homes)	• System size typically ranges from 1.5 to 4.0 kW, however 2.3kW is considered the "sweet spot" • Since new homes in California are often built to Tier 2 standards, which is 35% above California code, a 2.3kW system can cover 50-60% of a home's load • Increasingly, homebuilders are using building integrated PV products which improve aesthetics	• Installations generally range from 100-kWp to 1-MWp with systems as large as 10-MWp; recent trend has been toward larger systems (reduces transaction cost per watt installed) • For retail customers, PV system typically covers 30-50% of load; ranges widely for other customers from a trivial amount to 100%; limitations on system size include roof space and interconnection standards	• These larger systems are generally > 1MW size

Table 4-2. (Continued).

Leading Business Models			
	End-user Owner: Residential Retrofit	End-user Owner: Residential New Construction	Third-party Owner: Commercial Retrofit	Third-party Owner: Grid-sited
	• $ 9.50/Wac (2006); LCOE is still generally above residential utility rates, even with incentives • Interconnection with utility, net metering (available in most markets)	• Inverter size is a challenge for new homes; inverter manufactures are following the residential retrofit trend toward larger systems, which are oversized for most new home systems	• Use of storage and controls have been minimal, although remote monitoring is common • $ 6.25-$8.75/Wac; in some cases where the cost of electricity is high, systems can be cost	

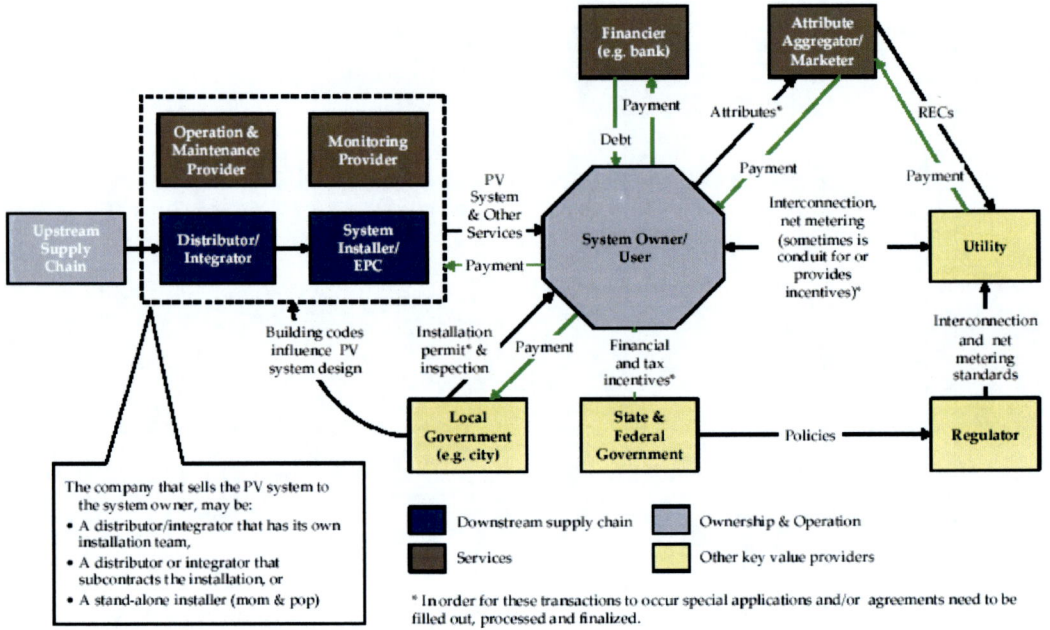

Figure 4-6. Value Network for the End-user Owned/Residential Retrofit model.

As would be expected, the value network changes for the other basic models examined in this section changes; these changes are discussed and shown below.

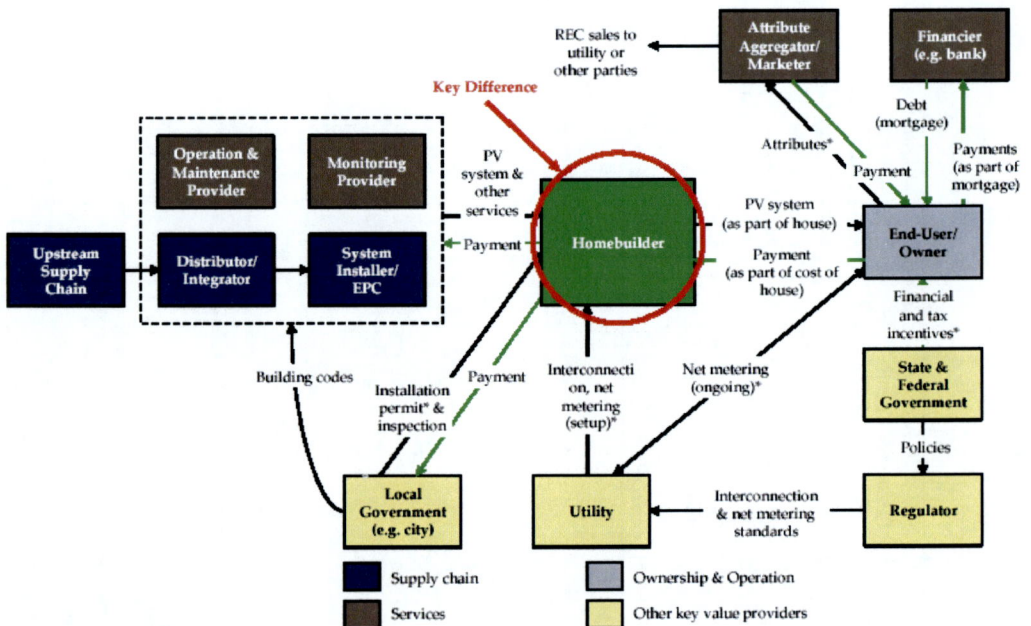

* In order for these transactions to occur special applications and/or agreements need to be filled out, processed and finalized.

Figure 4-7. Value Network for the End-user Owned/Residential New Construction model.

For the End-user Owned/Residential New Construction Model shown in Figure 4-7, the most striking difference from the previous model is the homebuilder as a central player in the installation of the PV system. Instead of the homeowner having ultimate responsibility for many aspects of the installation, it is the homebuilder that must generally manage interconnection with the utility and permits from the city. In this basic case, the homeowner still applies for the rebates and other incentives and manages any potential sales of RECs, and ultimately owns the PV system. There are variations to this basic model, which will be discussed later.

For the Third-party Owned/Commercial Retrofit model shown in Figure 4-8, the most striking difference with the previous model shown is the third-party as the central player, managing all aspects of the installation and then taking on the long-term ownership, operation, and maintenance of the system. The end-user is involved by way of providing roof space and purchasing the electricity (kWhs) that is generated from the system.

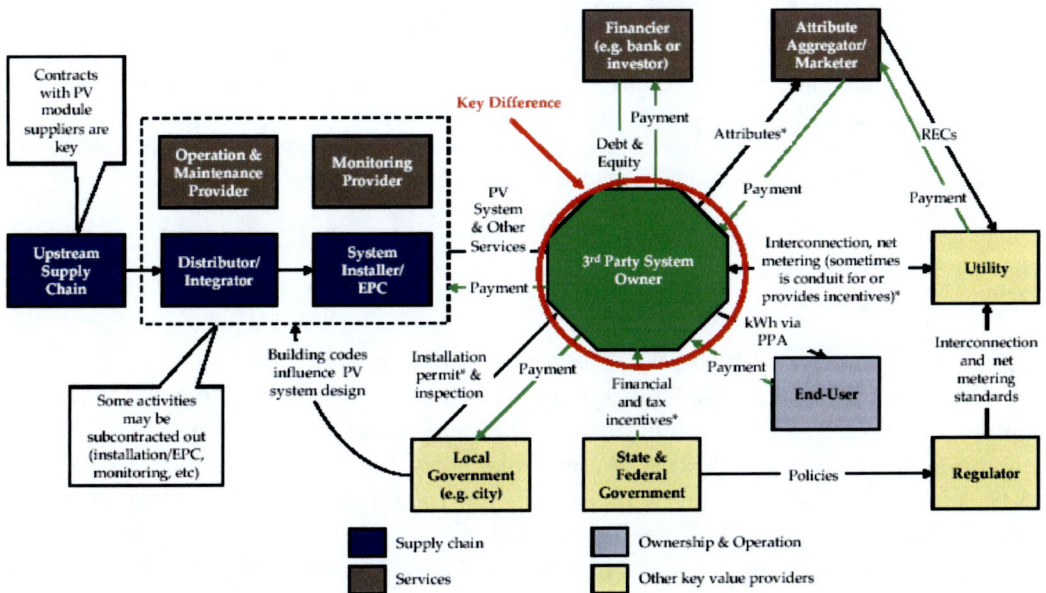

* In order for these transactions to occur special applications and/or agreements need to be filled out, processed and finalized.

Figure 4-8. Value Network for Third-party Owned/Commercial Retrofit model.

Finally, for the Third-party Owned/Grid-Sited model shown in Figure 4-9, the most striking difference from the previous model is that the relationship with the utility moves beyond interconnection and net metering, and now includes the sale of electricity (kWhs). This is a typical arrangement in the electricity sector where an independent generator establishes a purchase power agreement with a utility.

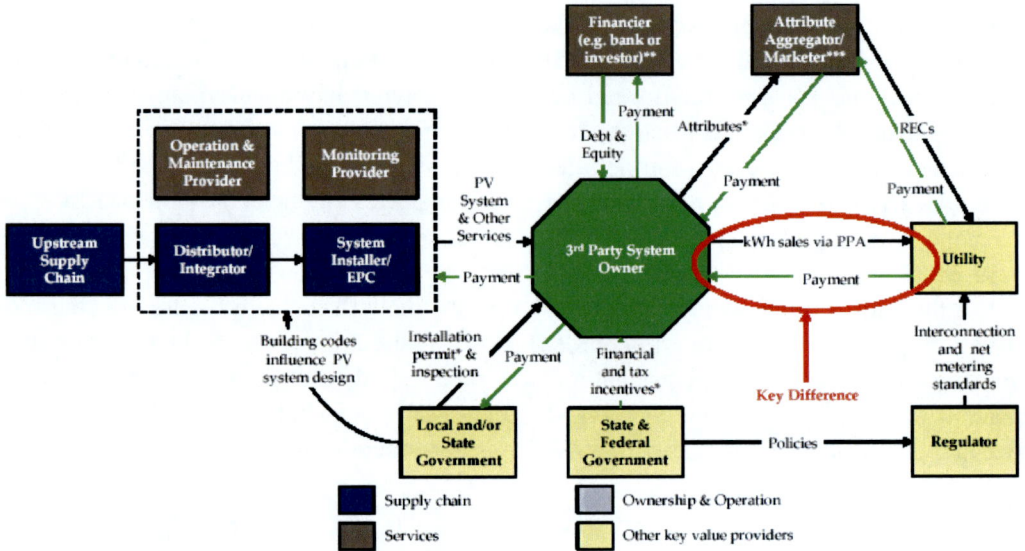

Figure 4-9. Value Network for Third-party Owned/Grid Sited model.

•In order for these transactions to occur special applications and/or agreements need to be filled out, processed and finalized.
** Access to financing could be from internal sources (e.g. the company's balance sheet or ex isting lines of credit). Project could include multiple investors.
*** Attributes could be sold directly to the utility.

4.2.3.3. Comparison of Business Model Variations and Trends

The business models characterized and shown in the sections above are intended to capture the most basic types. For each of the models described, variations exist in the marketplace. These variations allow companies to sell to different customer types (e.g., early adopter versus pioneer) and also demonstrate trends in the industry, such as reducing hassle and complexity for the customer or end-user. Table 4-2 summarizes current variations and trends by model type. Note that the Third-party Owned/Grid-Sited model is not included, as there are so few examples of this type in the United States, thus variations on the basic approach do not yet exist.

4.2.4. Utility Involvement in Current PV Business Models

Apart from involvement with net metering and interconnection, historically, utility participation in the PV market was limited to a few cases of retailing PV systems and providing system rebates. These later types of programs were generally undertaken by municipal utilities. However, a growing number of investor-owned utilities have recently taken on more active roles in encouraging the distributed customer-owned PV market. Using our taxonomy of PV business model evolution, this type of utility activity falls under *1st Generation* utility facilitation of PV business models. This activity is seen as supporting other business models, and not as stand alone business models. Table 4-3 below provides a brief description of current initiatives that utilities have created to facilitate the development of PV.

Table 4-3. Comparison of Business Model Variations and Trends

	Leading Business Models		
	End-user Owner: Residential Retrofit	End-user Owner: Residential New Construction	Third-party Owner: Commercial Retrofit
Business Model Variations	• Basic: The integrator/installer sells the system to the homeowner and installs it. Additional services might be O&M and system monitoring. The homeowner deals with all other aspects of the system installation, including: financing, utility interconnection, city permit/inspection and government incentives. Independent from the installation and system sale, banks provide financing for PV systems, but largely within their existing portfolio of products, such as home equity lines of credit.	• Basic: In the basic business model the homebuilder works with the installer or directly with the manufacturer to select the system and then manages all aspects of the installation, including permits with the city and interconnection with the utility. The system may initially be paid for by the homebuilder. The homeowner is responsible for processing the state and federal incentives, although the homebuilder or installer/integrator may help facilitate the process.	• Basic (e.g., "Hold"): A third-party owner orchestrates the entire process, acting first as project developer to identify project and system user, and then becoming system owner. In this model, the thirdparty owner raises capital (generally debt) to finance the company, holds the project and uses the tax incentives and accelerated depreciation itself. The third-party enters into a power purchase agreement (PPA) with the end-user. Typically 5 – 20 year contract terms with nominal performance guarantees. In this model, the third-party can be a stand-alone business (e.g., MMA Renewable Ventures) which hires a company to do the installation, supply the system and perform O&M and monitoring (e.g., SunPower and others), or alternatively the third-party can incorporate portions of the integrator, installer, O&M and monitoring businesses (e.g., SunEdison).
	• Hassle Free (almost): The integrator/installer takes on additional pieces of the installation process to streamline it for the customer. These include: interaction with the utility for the interconnection/net metering, application for city permit and attendance of inspection, application for government incentives, and rebates. The homeowners still pay for the entire system, arrange financing, and wait for incentives/rebates reimbursement.	• Roofing Company: In this model a roofing company becomes the integrator and installer, working directly with a PV manufacturer. The roofing company is subcontracted by the homebuilder to install all aspects of the roof, including the PV system. For roofing companies, this is potentially a lucrative source of new business (e.g., a solar roof is 4-5 times the cost of a conventional roof). Roofing companies understand and are able to manage the liabilities associated with roofing new homes.	• Equity and Tax Credit Driven: In this model, the thirdparty transfers partial or entire ownership to an equity investor (or group of investors via syndication) looking for tax and accelerated depreciation benefits. Typical investors are investment banks (e.g., Goldman Sachs) and also commercial banks and corporations. Project size is a consideration; these investors generally require projects greater than 2 MWp (around $ 25 million in size).

Table 4-3. (Continued).

Leading Business Models		
End-user Owner: Residential Retrofit	End-user Owner: Residential New Construction	Third-party Owner: Commercial Retrofit
• One-Stop-Shop: In addition to all of the services offered in the Hassle Free (almost) model, in this model the integrator/installer provides the customer with pre-arranged financing options from mainstream banks or financing entities (e.g., GE Capital Finance), and sometimes fronts the value of the state rebates for the customer, thus reducing the transaction costs and hassle associated with capital cost of the system. Homeowner still applies for federal tax credits.	• Standard Feature: In this model, the homebuilder includes solar PV as a standard feature of the home (e.g., Lennar Everything's Included Package).	• Lease: In this model, the third-party transfers ownership to a bank or leasing company (e.g., National City) which then offers lease financing to the user. This model is targeted at commercial end-users with large roof areas and a desire to access the benefits of solar. Projects tend to be smaller than Equity and Tax Credit Driven project, about 200kWp -2MWp
• Broker: A broker meets with the homeowner, determines their needs, and develops a request for quote, which is sent to local distributors/ integrators/installers (e.g., Sun Engineer in California).	• Zero Energy Home: In this model, the PV system is part of an approach used by the homebuilder, possibly in collaboration with the utility, to develop a home with minimal consumption from the grid. The home uses energy efficiency measures to reduce the home's consumption and solar PV to reduce the energy demanded from the grid.	• Build Own Operate Transfer (BOOT): A PPA can be structured so that after the tax incentives have been fully monetized, the system can be sold to the customer/user of the electricity. This generally occurs at year six of the PPA. This is analogous to the IPP BOOT model or flip structures used with wind power projects.
		• Tax Exempt Municipal Lease: Utilizes a municipal utility's ability to access low cost financing.

	Leading Business Models		
	End-user Owner: Residential Retrofit	End-user Owner: Residential New Construction	Third-party Owner: Commercial Retrofit
Key Business Model Trends	• Reduced hassle and complexity for customer • Increased marketplace competition and commoditization • Increased sophistication of back-office processes • Reduced customer acquisition costs • Increased use of remote monitoring, verification, and diagnostics	• Building integrated products are in greater demand • Roofing companies are entering the PV installation business • PV is becoming a standard feature in some new home development • PV is being considered with energy efficiency measures in design	• Larger systems especially commercial systems with PPAs, are tending toward larger sizes • Increased number of third-party intermediaries • Increased availability of capital

Table 4-4. Examples of Utility Programs
Supporting Current PV Business Models

Program	Utility	Brief Description
Financing	APS	Building a structure through which banks and lenders offer special financing or refinancing to solar customers (e.g., APS pays the PV rebate directly to the lender and the incentive is used to buy down the interest rate or to re-amortize the loan).
Financing	PSE&G	Proposing to lend capital to end-users and solar developers for 40%- 50% of the project cost, which is repaid over 15 years with S-RECs at a rate of 12.11%. S-RECs are valued at their floor price or current market price, whichever is higher.
Technology partnership	NSTAR	Aligning with Evergreen Solar to lower the overall cost of solar generation by promoting standardized systems installed by pre- approved solar contractors.
REC Database	Xcel Energy	Partnering with Pioneer Solutions to develop a software application to view and track RECs for compliance and trading purposes.
Feed-in Rate	We Energies	Offering a feed-in tariff of 22.5 ¢/kWh. Eligible PV systems must be between 1.5kW and 100kW, and customers must enroll for a 10-year contract.

4.2.4.1. Arizona Public Service: Financing

APS is building a structure through which banks and lenders offer special financing or refinancing to solar customers. For example, APS would pay the PV rebate directly to the lender and the incentive would be used to buy down the interest rate or to re-amortize the loan.

The goal of APS's financing program is to capture the consumer market that wants solar for the financial reasons, not simply because it is a good environmental choice. APS has done significant R&D to understand its customers' needs, and is taking a national leadership role amongst utilities that promote PV installation to customers.

Through this program, homeowners can understand the economic savings of PV as it is built into their mortgage. The lender accepts a deferred payment of the amount of the rebate from APS, as the mortgage is signed before the PV system is installed.

4.2.4.2. Public Service Energy Group: Financing[13]

Different from APS's program, Public Service Energy Groupd (PSE&G) would invest $100 million over two years to help finance the installation of solar systems for customers in its service territory. Capital would be lent to end-users and solar developers for 40%-50% of the project cost, which would be repaid over 15 years with S-RECs, at a rate of 12.11%. PSE&G would be repaid in S-RECs, the "currency" of the solar market in New Jersey, which are valued at their floor price or current market price, whichever is higher.

PSE&G's financing program is awaiting approval from the New Jersey Board of Public Utilities, but the initiative could fulfill up to 50% of the solar requirements for the New Jersey RPS in the utility's service area for the years 2009 and 2010, or 30 MW of capacity through 2010. The program would be paid for through funding currently earmarked for renewable energy programs.

---

[13] Not reviewed yet by PSE&G.

4.2.4.3. NSTAR: Technology Partnership

NSTAR has partnered with Evergreen Solar, the manufacturer of the low-cost String Ribbon solar panels. The program seeks to grow solar power installations, in response to customer demand, in the Eastern Massachusetts region (service territory of NSTAR). Through this partnership, NSTAR hopes to make solar installations more accessible and affordable by raising the awareness to its customers of solar options. The utility is exploring whether or not economies of scale can be gained in a vendor model similar to the existing energy efficiency process. For example, NSTAR contracts vendors to respond to customer home energy efficiency audit requests. It may be possible to provide the same service for solar-related home audit requests as solar becomes more prominent in the state.

By partnering with Evergreen, NSTAR can become more knowledgeable about solar and be better equipped to deliver on customers needs. The challenge in Massachusetts is that solar generation is still an expensive alternative to other forms of generation. There are rebate programs in place to help lower the cost and these programs will be critical in helping to develop the PV marketplace.

4.2.4.4. Xcel Energy: REC Database

Xcel Energy has worked with Pioneer Solutions to roll out a comprehensive database that creates a centralized, flexible system that tracks vintage, source, type, and initial value of each REC using first-in-first-out (FIFO) accounting. The initiative is driven by the fact that trading markets are slow to develop due to the absence of a real-time tracking system. Xcel Energy expects that the states in which it operates will now create REC markets its customers can participate in, along with Xcel.

The goal of the database is to track RECs throughout the utility's eight-state service territory and compile them in a database to manage RPS compliance and potential trading. Xcel Energy is looking to create value by creating a system through which "credible and easily audited REC transactions" can take place. Given that states around the U.S. continue to pass RPS requirements, Xcel Energy is positioning itself to capitalize on the development of future REC markets.

4.2.4.5. We Energies: Feed-in Rate

We Energies' feed-in rate is more similar to solar rebates seen in Europe. The utility offers a purchase rate of $0.225/kWh for 100% of the solar power generated, and the customer gets a bill credit or receives a check when the accumulated amount exceeds $100. A second meter is required for the program, which costs $2.50 per month for timeof-use customers and $1.00 per month for all other customers generating ≤ 40kW. Eligible PV systems must be between 1.5kW and 100kW, and customers must enroll for a 10-year contract and sign a standard interconnection agreement. Customers can leave the program with 60-days notice.

The feed-in rate is part of We Energies' Energy For Tomorrow "green pricing" program supply mix, which aims to increase Renewable Energy resource use for generating electric power and gives customers the choice to invest in helping the environment. The utility gets national recognition for being a leader in the renewable energy field and is able to increase customer satisfaction.

In addition to utility programs that help facilitate current business models, there are a few cases of utility ownership that could pave the way for greater involvement in distributed PV and a more active role either in ownership or control of PV systems (Table 4-4). This type of activity falls into our definition of *2nd Generation* business models as they include utility *ownership*.

Table 4-5. Examples of Utility PV Ownership

Progr am	Utilit y	Brie f Description
Customer- sited PV	SDG&E	As part of its sustainable communities program, the utility is installing PV systems at customer-sites on the utility side of the meter. The program was approved by regulators and SDG&E has $4.3 millio n per year. SDG&E is working with the LEED certifying body so that utility's PV systems can help the building owner achieve LEED status.
Solar Shares	SMUD	SMUD is developing a grid-sited 1MW PV system which would allow ratepayers to buy "shares" in it through a surcharge on monthly electric bills. The program aims to attract homeowners or commercial customers that want solar, but cannot install it because they rent, have shading issues, or do not have access to up-front capital.
Services Agreement (with ownership option)	Austin Energy	Considering a program to lease land to project developers. The developer would build a PV or concentrating solar system and then utilize/monetize the tax benefits, perhaps benefiting from Austin's access to low cost tax-exempt debt, after which the developer may have the option to transfer the ownership to Austin Energy. Additionally, Austin Energy is considering prepayment options as well as possible lease arrangements.

4.2.4.6. San Diego Gas & Electric: Customer-sited PV

San Diego Gas & Electric's (SDG&E) Sustainable Communities Program encourages sustainable building practices in the San Diego area while advancing the utility's electricity delivery system. The program:

1   Provides cash incentives for sustainable building projects,
2   Integrates utility-owned clean generation systems within sustainable buildings, and
3   Creates local showcases that serve as models for other projects.

In terms of siting new clean energy systems, the program works in the following way: systems are installed, owned, maintained, and operated by SDG&E. All design, installation, and maintenance work is contracted out. Participants "host" generation systems by leasing roof space to SDG&E, generally for a 10 year period, with two possible five-year extensions. The clean energy system is connected to the utility side of the meter and the electricity flows right into the grid; there is no net metering and no effect on the customer's electricity bill. The building owner can use presence of clean energy system in/on the building toward Leadership in Energy and Environmental Design (LEED) credits (following USGBC credit interpretation) and Collaborative for High Performance Schools (CHPS) points.

The program has a $4.3 million capital budget approved by regulators. The investments made under this program are rate-based and were submitted as part of electricity T&D testimony in the 2004 Cost of Service Filing. Clean energy systems (all PV to date) are strategically selected throughout SDG&E's service territory. Three installations were completed as of October 2007, with a total installed capacity of 190 kW. Over 1 MW is expected to be installed by the end of 2008.

4.2.4.7. Sacramento Municipal Utility District: Solar Shares

Under Sacramento Municipal Utility District's (SMUD) program, a solar contractor will build, own, and operate a 1 MW system within SMUD service territory. SMUD will buy all power produced by the system under a PPA and resell power to customers at a subsidized price. Instead of offering rebates for customer owned system, SMUD is reducing the price of the solar electricity. Customers will buy blocks of power from the system (e.g., representing .5 or 1 kW of capacity) for a fixed monthly price. The customer will have a "virtual PV system" that acts like a customer-owned system such that the customer will have net billing with kWh generation matching output of real system and a fixed monthly payment, similar to financing a rooftop system. SMUD estimates that it can service 900 customers with this offering (assuming a mix of small, medium, and large customers).

The goal of the Solar Shares program is to makes solar available to all customers, including those that cannot or choose not to own their own PV system. For the utility, the program helps SMUD meet aggressive RPS and solar requirements by connecting more solar to the grid for less utility investment compared to customer-sited PV systems— economies of scale lowers installation cost and federal tax credits are maximized since solar developers can take full 30% versus $2k cap for residential customers. Finally, it demonstrates a sustainable business model that can be expanded in the future and emulated elsewhere, ultimately accelerating PV deployment by reducing costs and developing new markets.

4.2.4.8. Austin Energy: Services Agreement

Similar to other utility programs mentioned above, Austin Energy's services agreement approach is in its planning stages. The idea is similar to a relatively new structure in the wind industry, referred to as a "services agreement", which sometimes involves pre-payment for future power deliveries as a way to tap into the municipality's low-cost tax- exempt debt, as well as the private developer's access to federal tax benefits. In this case, Austin may lease the land to a developer and has the opportunity to own the solar system in the future. The developer would build a PV or concentrating solar system and then utilize the tax benefits, after which the developer would have the option to transfer the ownership to Austin Energy.

This business model helps Austin Energy meet the state's RPS requirement, while facilitating third-party ownership of solar projects. Austin Energy is still considering prepayment options as well as possible lease arrangements.

4.2.5. Innovative State Programs Encouraging Current PV Business Models

The final piece in this section briefly examines what states have done to accelerate PV adoption and new business model creation. While the majority of states require net metering and interconnection and some offer financial incentives for solar investment, a handful of states are creating additional programs and/or mandates to further stimulate PV market growth. States see the PV industry as a driver of economic growth and energy stability, and while only a few programs exist thus far, this could be an indication of growing support for PV at the state-level. Table 4- illustrates the initiative a few states have taken to promote solar development.

Table 4-6. Innovative State PV Programs

California	A new home builder mandate is part of SB1, the Million Solar Roofs bill. The bill requires that all builders of single-family home developments (>50 homes) offer solar power as an option.
Nevada	In 1995, the Federal government established a Solar Enterprise Zone (SEZ) in an area of southern NV. The DOE identified the SEZ as a preferred civilian use of the area that includes the former Nevada Test Site. The ultimate goal was to deliver 100MW of solar power to the state.
New Jersey	A solar REC market was developed as part of the state's Clean Energy Program. Individuals and businesses can finance their solar installations by trading an S-REC once the system generates 1 MWh. Utilities can purchase S-RECs to meet the required RPS.
Washington	Solar PV Feed Law pays 15¢/kWh for electricity generated by solar panels, and if the panels are manufactured in Washington, a 54¢/kWh payment.

4.2.6. Observations

The current and emerging business models discussed in this section almost completely fall within the *0* and *1ˢᵗ Generation* of PV business model evolution. Much of the innovation and change taking place with current business models is concerned with creating a more efficient, streamlined, and mature delivery channel to the customer. These are important steps in widening the customer base and driving down cost.

As this study progressed, it became apparent that the ownership attribute remained a key factor in defining future business models, but that the characteristic of application was not critical. Instead, the question of who controls the PV system became a defining factor of future business models; thus, future business models are characterized in the next section by ownership and control. As discussed in the Future Business Models section below, we assume that as the PV supply chain matures, the application (e.g., where the system is installed) will become less important than who controls the system and how.

The next section on Future Business Models discusses the evolution to *2ⁿᵈ Generation* business models that more fully incorporate PV as part of the electric supply and distribution infrastructure. One of the significant gaps between the *1ˢᵗ* and *2ⁿᵈ Generation* business models, which will be explored in the next section, is the use of advanced controls and combining PV with other distributed resources and loads. As shown in this section, this type of activity is almost exempt from today's business models. In addition, while utility control and ownership is negligible today, it is explored in the next section.

4.3. Future Business Models

4.3.1. A Long-term Vision for PV

In developing future business models for PV, it is necessary to envision the long-term future for the PV marketplace. This helps to identify business models that will be:

- Successful in growing PV markets in the near to medium term, and
- Most capable of addressing the challenges that will arise from increased market penetration of PV.

Looking forward 10 to 20 years, there is a strong case to be made that PV in distributed applications, primarily customer-sited, will become an inevitable and significant component of the electricity sector, especially if forecasted PV cost reductions materialize. The long-term vision is one where PV has passed a "tipping point", beyond which PV is considered competitive with retail power supplied by the grid (Figure 4-10). PV will be competitive when a large segment of electric customers considers the overall value proposition from PV to be competitive with grid power. This will move PV customers beyond so-called *early adopters* into what is sometimes called the *early majority,* a large segment of consumers that adopt the technology during a period of rapid market penetration.[14] PV may also become competitive for large-scale, centrally-sited PV (e.g., grid-sited), but in this vision, the focus is on distributed PV, where it has the greatest value and greatest implications for grid operations.[15] For PV to be competitive, it is not necessary for it to have a levelized cost of electricity below the average retail grid power price or even a time-of-use price. There are other attributes for which there is demand and for which there is increasingly real economic value, such as carbon avoidance, other environmental benefits, and increased reliability. New business models can help unlock this value.

Reaching the point of wide-scale competitiveness with grid power may come sooner, as a result of specific breakthroughs in technology, or later, as a result of the steady march down the cost curve. In either case, it is assumed that the PV supply chain is able to ramp-up capacity to supply the market demand.

Because the resource potential for PV is effectively unlimited (i.e., compared to electricity demand), and PV can be sited virtually anywhere and at any scale, the long-term vision for PV is one in which PV is a major component of electricity supply. Moreover, as the benefits to society and ratepayers of PV deployment become greater, policymakers and utility regulators can be expected to take action to facilitate widespread adoption of the technology. This has significant implications for key stakeholders, especially the utility and the end-user. In unbundled power markets, wholesale power suppliers and transmission companies may also be impacted in a meaningful way as more and more demand is met with distributed resources, which could lead to stagnant, if not decreasing, demand for wholesale power and related transmission service.

Table 4-5 summarizes the main implications to key stakeholders of such a long-term vision. Consistent with the overall approach in this chapter, this section focuses on the ownership and operation of the PV systems, and not on the PV product supply chain. Over time, the PV equipment supply chain can be expected to look like other equipment supply industries (emergence of a few large suppliers, similar margins, ready access to financing, standard product offerings, and overall competitive structure). Conversely, stakeholders that are involved in the ownership and operation of PV systems will be participating in a major reconfiguration of the traditional power supply business. This is where innovation and disruption in "business as usual" will occur, and where there is potential to unlock the value of PV.

---

[14] Geoffrey Moore, *Crossing the Chasm*, HarperBusiness, 1991.

[15] Large-scale PV systems deployed on the grid will, in the absence of energy storage, behave a lot like peaking plants, and are not the focus of this analysis.

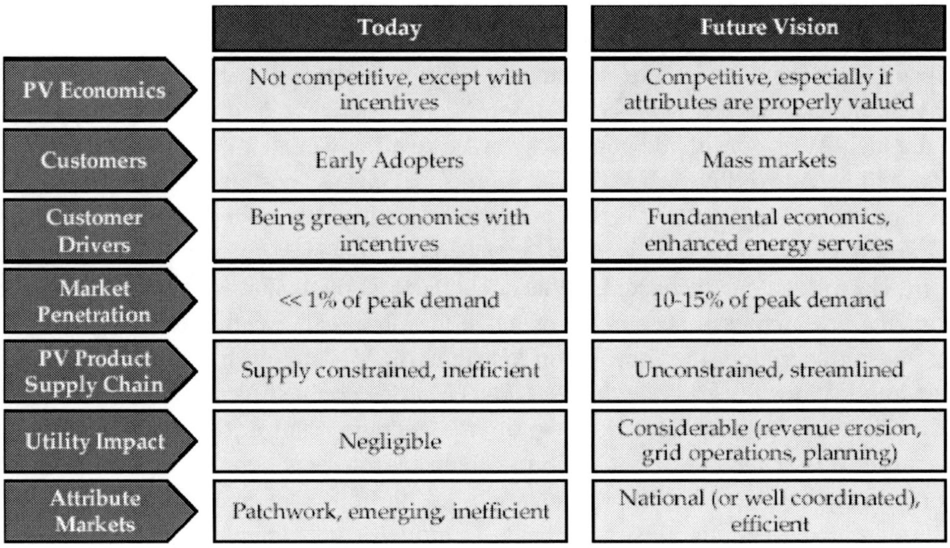

	Today	Future Vision
PV Economics	Not competitive, except with incentives	Competitive, especially if attributes are properly valued
Customers	Early Adopters	Mass markets
Customer Drivers	Being green, economics with incentives	Fundamental economics, enhanced energy services
Market Penetration	$\ll 1\%$ of peak demand	10-15% of peak demand
PV Product Supply Chain	Supply constrained, inefficient	Unconstrained, streamlined
Utility Impact	Negligible	Considerable (revenue erosion, grid operations, planning)
Attribute Markets	Patchwork, emerging, inefficient	National (or well coordinated), efficient

Figure 4-10. A future vision for PV.

Table 4-7. Implications of Widespread Distributed PV Deployment on Key Stakeholders

Stakeholder	Implications
	PV system: <ul><li>Is cost-effective alternative to the grid</li><li>Provides improved reliability (over grid)</li><li>Helps meet environmental desires of consumers</li></ul>
End-user	<ul><li>Generates a range of value streams (driven in part by environmental and climate change policy)</li><li>Is part of a bundle of new technologies to improve energy service at end-use and reduce cost as cost drops (low-cost energy storage, distribution system automation, "smart homes", plug-in hybrid vehicles)</li></ul>
Stakeholder	Implications
System Owner	<ul><li>PV system output has multiple value streams that can make it competitive in the market relative to grid power</li><li>Owner needs to be able to identify and capture multiple PV value streams</li></ul>
Distribution Utility and Vertically Integrated Utility	<ul><li>High degree of PV market penetration creates:</li></ul> o Reduced throughput leading to revenue loss under traditional tariff structures o Need for control of PV systems and/or new distribution system architectures to ensure safety, operational integrity and reliability of the distribution grid<ul><li>In addition, new technologies used in conjunction with PV could radically change utility operations and product/service offerings to customers (low-cost energy storage, distribution system automation, "smart homes", plug-in hybrid vehicles)</li></ul>

Wholesale Generator	• High degree of PV market penetration could provide competition in the wholesale market to more expensive generating assets
Regulator	• Emergence of cost-effective PV and other complementary technologies creates need for major transformation of how utility industry is regulated
Transmission Company	• High degree of PV market penetration could impact the demand for transmission services

4.3.2. Other Developments and Considerations

At the same time that the PV industry is making great strides in the deployment of PV using *0* and *1st Generation* approaches, significant activities are also occurring outside of the PV industry that have clear implications for long-term PV market penetration. In particular, changes in policy, technology and utility regulation may hold the potential to not only create opportunities to unlock additional value from PV systems, but may simultaneously create more demand for it (Figure 4-11).

Technology developments underway to manage the distribution grid more effectively will have many benefits for distributed generation, including PV. In particular, the development of distribution system automation, the transition to "smart grids," and the deployment of customer- and utility-controlled demand response are all likely to help utilities and others unlock additional value from distributed PV systems.

Policy trends that create a market for renewable energy, such as Renewable Portfolio Standards or RPS (especially those with solar set-asides) and greenhouse gas emission caps, are gaining momentum at the state and local levels, and may ultimately culminate in much higher average state targets and, eventually, a federal-level policy.

Finally, regulatory changes in some states are altering the way a utility perceives its business. Beyond net metering and interconnection issues, performance-based ratemaking (in which incentive benchmarks, rather than budgets, determine cost recovery) and revenue decoupling mechanisms (in which rates are determined as a function of service delivery rather than as a strict return on hard assets) are being implemented to encourage energy efficiency, conservation, and renewable energy. Given these types of changes, the ability of a utility to realize revenue from rates that are based in part on reconfiguring its grid and altering its customer support to integrate PV will have obvious benefits for the further increase of distributed PV. In addition, some utilities have experimented with tariff structures to encourage desired consumer behaviors and the deployment of new technologies. For example, variations of time-of-use pricing can be very beneficial to PV economics. Also, adoption of transmission congestion pricing should have a beneficial impact on distributed PV, as the market value of distributed generation will be made plain by the congestion prices. These regulatory actions are increasingly being driven by the desire to encourage conservation or greenhouse gas reductions.

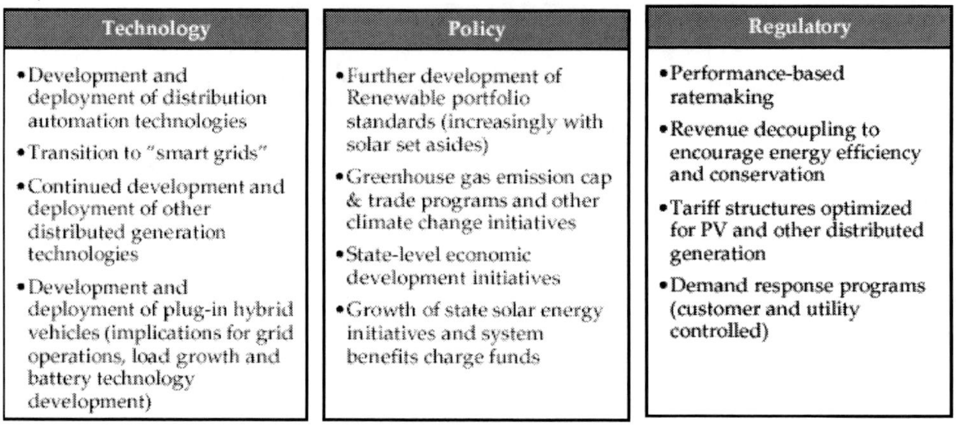

Technology	Policy	Regulatory
• Development and deployment of distribution automation technologies • Transition to "smart grids" • Continued development and deployment of other distributed generation technologies • Development and deployment of plug-in hybrid vehicles (implications for grid operations, load growth and battery technology development)	• Further development of Renewable portfolio standards (increasingly with solar set asides) • Greenhouse gas emission cap & trade programs and other climate change initiatives • State-level economic development initiatives • Growth of state solar energy initiatives and system benefits charge funds	• Performance-based ratemaking • Revenue decoupling to encourage energy efficiency and conservation • Tariff structures optimized for PV and other distributed generation • Demand response programs (customer and utility controlled)

Figure 4-11. Ongoing energy sector developments affecting distributed PV markets.

4.3.3. Utility Structure and Core Competencies

To understand how electric utilities will become more involved in distributed PV as a business opportunity, it is important to review the basic structure of the electric utility industry and to understand utility core competencies. Figure 4-12 shows a basic electric utility value chain, from fuel supply to end-user. The picture is complicated because some states have unbundled (restructured) their electric utilities, and in these instances, each of the basic functions shown in Figure 4-12 is performed by separate companies.

For the purposes of this chapter, the key element to consider is that restructuring has separated the basic functions of generation, distribution, and retail power sales (e.g., transmission may still be owned by distribution companies). In these cases, the companies that own the wires no longer own generation, and the retail sale of electricity is opened up to competition. The distribution (wires) companies, which are still subject to state regulation, also serve as "default service providers", selling electricity to customers that do not switch to competitive energy suppliers.

Large municipal utilities (i.e., that own generation) and vertically integrated investor-owned utilities (IOUs) still control the entire value chain. As such, they own a significant portion of the capacity used to meet their load. These assets are in the "rate base". However, even with vertically integrated IOUs, the wholesale power markets are competitive in that IOUs cannot simply add generating capacity to meet their demand growth. Instead, they must competitively procure this capacity. In this case, the power purchase agreements are typically treated as recoverable cost and thus utilities are not permitted to earn a return.

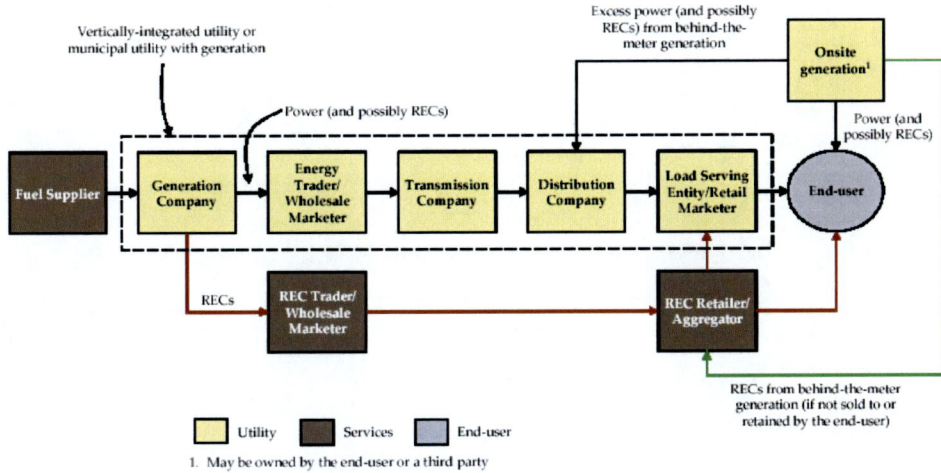

Figure 4-12. Simplified electric power value chain.

IOUs are subject to regulation by the state utility commission, whereas municipal utilities generally are not. In restructured markets, utility regulators generally focus on the distribution companies, while wholesale generation companies and load serving entities are generally considered unregulated businesses. Smaller municipal utilities (i.e., the majority of municipal utilities that do not own generation) resemble distribution companies but also act as the load serving entity.

Unbundled power markets present some unique issues for distributed PV, in particular, how different value streams can be captured when there are several different entities along the electric power value chain. For example, since a distribution company does not generally own generation, value that PV has for offsetting the need for additional fuel or generation capacity, is not easily captured. Conversely, to the extent that the value of PV can ripple up the value chain, a vertically integrated entity can internalize all of those benefits. Also, distribution companies in unbundled markets may be barred from owning generation, but future PV business models may center around utility ownership of PV. Thus, regulatory changes may be needed to address this and other issues.

The previous discussion is important because it has implications for the regulatory changes that may be required as utilities become more deeply involved in the PV market. Nevertheless, whether talking about vertically integrated vs. unbundled markets, or IOUs vs. municipal utilities, the main parts of the value chain that will be impacted are distribution and retails sales. Thus, when this chapter refers to "utilities", it is primarily with these functions in mind.

Under the current regulatory structure, utilities (and other stakeholders in the electric power value chain) have a lot to lose from high levels of distributed PV market penetration. Nevertheless, utilities possess a range of core competencies and attributes that they can leverage for new PV business models. While they can vary depending on the type of utility, as described above, they generally fall into three categories (see Figure 4-13): asset management and investment, customer service, and system operation.

So while distributed PV may initially be viewed as a threat, it is also clearly an opportunity, and this is expected to be the central issue going forward as new PV business models are developed.

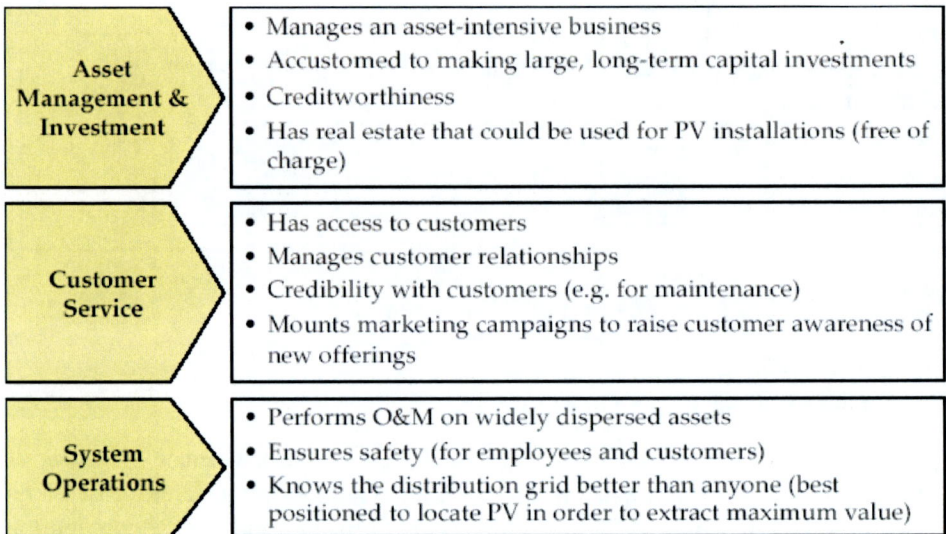

Asset Management & Investment
- Manages an asset-intensive business
- Accustomed to making large, long-term capital investments
- Creditworthiness
- Has real estate that could be used for PV installations (free of charge)

Customer Service
- Has access to customers
- Manages customer relationships
- Credibility with customers (e.g. for maintenance)
- Mounts marketing campaigns to raise customer awareness of new offerings

System Operations
- Performs O&M on widely dispersed assets
- Ensures safety (for employees and customers)
- Knows the distribution grid better than anyone (best positioned to locate PV in order to extract maximum value)

Figure 4-13. Utility core competencies and attributes to leverage for new PV business models.

4.3.4. Maturation of the PV Value Network

As discussed in Section 4.2, the emphasis in PV markets up to now has been largely on the PV product supply chain, and more recently, on reducing hassle for customers and third-party ownership of customer-sited PV. These areas continue to be the main focus of investment, as the industry attains large-scale status in manufacturing, product delivery, and installation. Over the next 5-10 years, one can expect these aspects of the PV value network to continue to mature and consolidate, and therefore, the PV product supply chain should come to resemble other businesses that supply key building infrastructure, like HVAC equipment. While there will still be room for customization, it is expected that to a large extent, the PV product supply chain will become more streamlined and efficient, and PV products will be more standardized. Similarly, other aspects of the PV value network will become "commoditized", such as financing.

Table 4-6 summarizes the expected evolution of the various elements of the PV value network.

Table 4-8. Maturation of the PV Value Network

Value Network Element	Trends
Upstream PV product Supply	• Increased scale of manufacturing to bring down costs • Improved ability to incorporate product innovations and improvements into standardized products • Supply chain will consolidate and become more efficient
Integrator/ Installer/EPC	• Supply chain will consolidate and become more efficient • High-quality installation will become common and fairly non-specialized (i.e., similar to current HVAC or roofing industry)
O&M	• Non-specialized firms will be able to provide O&M
Monitoring	• System monitoring will become highly automated and integrated with O&M provision and become integral to the provision of other services around the PV system and other energy systems and services
Financier	• Access to financing will be easy and will leverage existing channels • Less financing will be required per system given drop in system prices
Attribute Aggregator/Marketer	• As attribute markets mature, aggregator/marketers will grow in size and market will become more liquid and efficient, to resemble other commodity markets • This service may increasingly be bundled with other energy products and services, such as financing, energy management and energy supply • Large uncertainty is the pace at which the patchwork nature of the market today will grow and mature to maximize value extraction from attributes.
Project Developer	• For large projects there will still be a role for project developers, analogous to role played by wind farm developers today • Pure-play developers may not be common as this activity becomes incorporated with other elements of the PV value network, such as installation • For customer-sited projects, the role may evolve into one of a broker, whereby the broker works with a customer to solicit and evaluate bids from prospective installers (similar to a real estate agent).
Utility Regulator	• May realize benefit to ratepayers of more widespread PV and take action to facilitate widespread adoption • Inclusion of PV in rate cases and regulatory hearings will be commonplace
Local Government	• May have familiarity with applications for PV permits and will have made adjustments to permit process and building code to facilitate consumer demand (e.g., standardization across jurisdictions)
State & Federal Government	• Will reduce incentives considerably given the reduction in system price • Remaining incentives are likely to reflect environmental benefits • Mechanisms to monetize benefits (e.g., attribute markets) will be developed in favor of high-cost financial incentives such as rebates
Utility	• Increased understanding of and experience with the technology • See table 4-5

We assume that as the PV supply chain matures, the application (e.g., where the system is installed) will become less important as a defining characteristic of business models than who owns *and* controls the system. Thus in the new business models section below, the ownership attribute remained a key factor in defining future business models, but the characteristic of application is not longer used, as in the current business models section. Instead the question of who controls the PV system became a defining factor of future business models, thus, future business models are characterized in the next section by ownership and control.

4.3.5. New PV Business Models

New business models will become necessary when distributed PV becomes a large enough fraction of total load on the grid, such that it cannot be ignored from the perspective of grid management and utility revenue erosion. As such, new business models will be focused on greater integration and control of PV, potentially bundled with other value-added services that may be enabled by so-called "smart grid" technologies.[16] What will distinguish these new business models from current ones are:

- PV system ownership

  o Who pays for the system
  o Who receives the value generated by the system

- PV system operation and control

  o How to ensure safe, reliable and efficient operation of the grid
  o How to maximize value from the PV asset(s)

Of all the stakeholders involved, it is the utility that will have its existing business model disrupted the most, and must therefore adapt its current business model in order to protect and enhance its business. Thus, greater utility involvement is seen as the key to new PV business models.

Three basic types of business models are examined for the future (Figure 4-14), depending on who owns and controls the system. A fourth option, in which the system is owned by the utility but not controlled by the utility, is not consider a viable business model. As will be discussed in more detail below, the success of any of these types of new PV business models will be tightly linked to ongoing technology and market developments in distribution automation and demand response, and may also require significant regulatory changes.

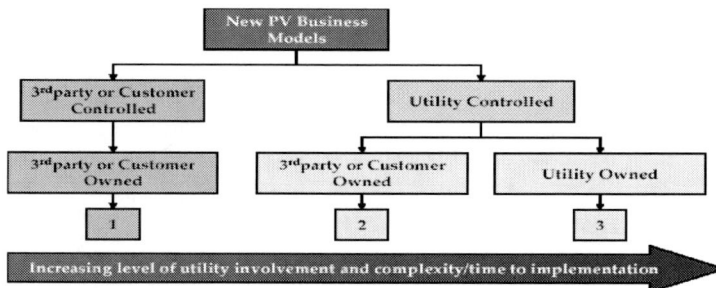

Figure 4-14. New PV business models focused on system ownership and control.

---

[16] For example, on September 26, 2007, FPL Group announced the launch of an initiative to invest up to $500 million to create a smart network that will provide its 4.5 million customers with enhanced energy management capabilities. This new program is designed to allow customers to view their energy consumption online every day, in real time, and to enable FPL to develop better energy management programs (FPL Group press release: *FPL Group Plans to Boost U.S. Solar Energy Production.* JUNO BEACH, Fla., Sep 26, 2007).

From a business model perspective, the combinations above need to be evaluated from two basic positions:

- What are the sources of revenue that support the business model
- What cost savings support the business model

For example, to the extent that there are system-wide benefits, these can generally be thought of as cost savings to the utility (e.g., reduced fuel consumption at central plants due to PV generation, the use of PV on-peak to make the T&D system more efficient, avoided investments in traditional generation assets). Other benefits may accrue to, or have value for, the end-user. For example, the environmental attributes of PV generation can be sold, producing revenue for the owner of those attributes, or if PV can be made part of a backup system, this service can have value to the end-user.

Depending on the type of business model and regulatory structure, dollars may flow from the utility to the owner/end-user, or vice versa. As highlighted in the DOE Renewable System Interconnection PV Value study, it is worth noting that the largest potential sources of value relate to central generation (both in terms of marginal generation costs and avoided capacity investment costs) and avoided investments in transmission and distribution. The next most valuable are the environmental attributes, driven by policies and regulations to curb criteria pollutants and greenhouse gases. However, the dynamic benefits of distributed PV to the grid (e.g., ancillary services, system losses, system resiliency) are much lower. This suggests that business models built around these lower- valued PV system attributes may not be viable, unless they can also take advantage of the other more lucrative value streams.

4.3.5.1. Third-party/Customer Owned and Controlled PV Business Model

4.3.5.1.1. General Description

In this business model, the customer or a third party controls the PV system as well as owns it. There is also the possibility of customer ownership combined with third-party control. This business model is primarily an extrapolation of current business models and trends (Figure 4-15). The key difference is that additional sources of revenue are captured by the owner, based on various changes to the regulatory and policy regimes, and on the deployment of "smart grid" technologies and energy storage that is integrated with PV system operation.

In this model, the utility role remains mainly one of facilitation—primarily driven by regulatory or policy changes; for example, in the increased use of demand response programs and greater implementation of S-REC markets, like in New Jersey. In this model, the utility pays for value-added products and services obtained from the PV system and is then allowed to recover these costs through traditional rate-making proceedings. In order to encourage this new model, regulators could consider allowing slightly higher rates of return—at least initially—as utilities adopt new practices.

Because an end-user with customer-side PV can generally already take advantage of net metering, the incremental value is expected to be modest. As such, this business model is not driven by the value of additional products and services. Rather, the growth of the PV installed base presents an opportunity for taking advantage of these assets. To the extent that there can be pricing or other signals that promote adoption of PV on certain parts of the grid (e.g., areas with constraints), this can serve to generate the most value possible.

This business model is considered the most likely to become established in the absence of outside influence, as various pieces of current regulation and policy are already in place to enable it in some jurisdictions. For example, PV with battery storage and/or in combination with load management would be able to fit under a demand response program, adding value over just net metering alone. Similarly, S-REC obligations in some states have already established a value for PV's environmental attributes. In addition, the incremental value provided by the PV system in this model is not dependent on having large amounts of distributed PV available. However, some values would certainly not be captured until additional regulations are in place.

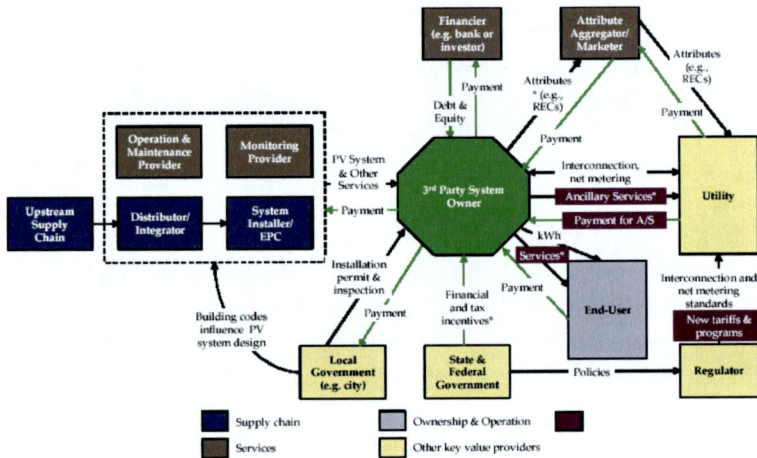

* Requires communications and control, including for performance[ ]based incentives.

Figure 4-15. Third-party/Customer Owned and Controlled Value Network.

The basic model shown above can be altered to show end-user ownership and control by simply replacing the third-party system owner with the end-user, and eliminating the final transaction for kWh sales. In addition, the supply chain and other services—including O&M, monitoring, financing and attribute aggregation—are shown separately for consistency with Sections 4.1 and 4.2. However, as discussed earlier, they could be combined in numerous ways, including provision of financing via the integrator or installer. In a similar way, the function of aggregating the attributes generated by the PV system could be performed directly by the utility.

4.3.5.1.2 Business Model Requirements

In this structure, the dollars are expected to flow mainly from the utility to the owner, with the regulatory structure making accommodations for this. In this sense the utility's business model would remain largely unchanged, as these transactions would be subject to cost recovery. In some variations (see table 4-9), activities may result in flows of dollars to the utilities (e.g., for maintenance or distribution wheeling charges). At the macro level, these business models will emerge in cases where overall, the current regulatory structure is maintained, but where the utility is made whole due to lost revenues, and where markets are created for attributes and there is potential allowance of distribution wheeling. It is notable that depending on how regulation plays out (e.g., if utility is made whole or not), this group of business models could lead to significant overall revenue erosion for utilities, especially as it results in a decrease in generation and throughput over the system.

Energy storage could further enhance the value of PV to both the utility and the customer. For the customer to benefit directly (i.e., PV as backup), current interconnection standards will need to be changed to allow for the system to continue to operate during a network outage.

Tariffs optimized for PV will also allow the end-user to maximize opportunities for cost savings. These could include time-of-use pricing, real-time pricing, enhanced net metering tariffs, and tariffs that would potentially allow for wheeling of excess generation across the distribution system. Changes to distribution pricing (e.g., locational pricing) may also send signals to end-users to encourage adoption of PV in areas of highest value to the utility.

As the market penetration of PV grows, utilities may be required to raise net metering caps or remove them altogether. Eventually, this may require some reconfiguration of the distribution system to accommodate the high penetration levels of PV. Furthermore, if PV systems are integrated with demand response, and utilities come to rely on this available capacity, utilities will likely seek certain commitments from demand response customers. Thus, it may be necessary to integrate PV systems with other load management options, including demand restrictions, should the customer be unable to meet their commitments.

Since the PV asset is neither owned nor controlled by the utility, the ability to send the appropriate price or other signals to many owners is important to ensuring the fleet of PV systems is of value to the utility. For this, some key technologies include:

- Advanced communications, monitoring and metering, which allow for response to pricing and other signals to maximize value to the end-user. These technologies could allow response to pricing signals for various attributes and could be used to verify system operation and delivery of network services;
- Communications and control technologies that allow third-parties to aggregate PV system output and actively manage systems as a fleet in response to various pieces of information from the grid; and
- Energy storage to firm-up capacity and allow for dispatch of the system, and to enable services to end-users.

Finally, even though the utility does not have control over the PV system in this business model, it might require the right to "see" the location and generation of the system on the grid, as the PV systems operation can impact overall grid operations. Advanced communications, monitoring, and metering technologies already in place could include two-way communication capabilities to provide utilities with basic system operation information.

4.3.5.1.3. Variations

A number of variations on the basic business model described here are possible (see table 4-7).

Table 4-9. Business Model Variations for Third-party/Customer Owned and Controlled PV

Business Model Variation	Description
Forward REC purchases	In active REC markets for RPS compliance, especially those with solar set- asides, forward purchases of REC streams could emerge as a viable business model. Utilities or other entities would cover all or part of the up- front cost of the system in exchange for a future REC stream. Forward REC purchase might be able to be rate based. Note: PSE&G has filed a concept similar to this with regulators.
Distribution-level ancillary services market	With the right regulatory and technological changes, a market for ancillary services at the distribution level could develop. The utility would pay for such services, and as the market develops, the utility could grow to rely on it and incorporate it into system planning. The utility would seek out the least-cost approach to procuring the ancillary services it needs to operate the distribution system. Demand response and VAR support are two examples.
Customer-focus	PV could become part of a technology package offered to end-users by third parties, such as energy service companies. PV could be integrated with energy efficiency performance contracting, energy procurement (in restructured markets), and building energy management services, including backup generation.
Distribution utility as network services provider	This vision of a distribution utility is most likely to emerge in unbundled markets. In this model, the utility (as the wires company) provides a range of network services to end users. This would require the decoupling of utility revenues from energy sales (this could include distribution wheeling). This model of the utility, if it emerges, could further enable the "customer focus" business model. Net-metering tariffs would be replaced by the ability of end- users or third-party owners to sell excess power over the grid to other end- users.

4.3.5.1.4. Challenges

As discussed above, the value of distributed PV to the operation of the distribution system is expected to be modest. As such, it may be difficult to create new business models centered around these values. Instead, these value streams are likely to be added incrementally to existing business models. Moreover, value streams from distributed PV can be broadly grouped into two categories—operations and planning. In these business models, the utility does not control the timing or siting of PV capacity; therefore, it is difficult for the utility to incorporate deployment of PV into distribution system capital planning, as the ultimate decisions of when and where to site are out of their control. This may result in a sub-optimal grid configuration and will make it more difficult for the utility to incorporate PV into capital planning activities.

In order to at least partially compensate for this issue, the utility could issue RFPs for network services. This would encourage siting in the best locations. The utility may also provide incentives to site in constrained areas to maximize grid value.

In unbundled markets, there are few ways to address the potential impacts of widespread adoption of PV on retail energy suppliers, wholesale generators, and transmission companies. These companies, however, could take on the third-party ownership role, thereby mitigating impacts on their wholesale business. Still, in the case of very high PV deployment, the potential exists for there to be stranded generation assets, in both vertically integrated and restructured markets.

If the customer/third-party controlled and owned business model becomes widespread, the distribution grid must be re-engineered to be highly responsive to changes in PV operating profiles (e.g., extremely localized power fluctuations), either due to transient changes in sunlight availability or to decisions taken by the owners, because the utility will not control the PV systems. An issue that will arise is the degree to which owners will be "free to choose" how to operate their systems. For example, if a customer chooses to participate in a demand response program, they might be obligated to respond to utility signals.

4.3.5.2. Third-party or Customer Owned and Utility Controlled PV Business Model

4.3.5.2.1. General Description
This business model is somewhat similar to those described above, in that they seek to achieve similar objectives (Figure 4-16). The key difference is that greater utility involvement in the operation and control of the systems is contemplated as a way to increase the value of the assets. Like the customer controlled business model described above, regulatory and policy regimes will need to change—though more significantly here—to allow the utility to reach behind the meter where the PV system will reside. In this case, the customer will not respond to price signals because the utility is controlling the PV system, at least to some extent. This business model will also require the deployment of "smart grid" technologies and would be enhanced by energy storage.

This business model may work best where aggressive demand response or other similar programs are being pursued or where high penetration of PV systems may pose serious grid control and operations issues. Under those circumstances, direct utility control—for example, to allow the utility to curtail PV system operation to maintain grid stability— instead of a complicated market for such services, may be preferable because the utility is assured response as it controls the asset as opposed to relying on optional response to price signals.

In this model, the utility would still pay for value-added products and services, and would then be then allowed to recover these costs through traditional rate-making proceedings. To the extent that PV systems provide a service and create value (e.g., avoid costs) for the utility, this would be factored into the cost of recovery calculation

This business model is expected to evolve more slowly given the additional regulatory changes required to permit utility control behind the meter. Additionally, distributed PV needs to exist at a significant scale in order for a utility to find value in controlling it. For example, the distributed PV installation would have value to the utility proportional to its capacity to substitute for generation, capacity, and transmission and distribution (T&D) investments.

As discussed in the section above, this diagram could represent the end-user as the owner by simply replacing the third-party system owner with the end-user, and disregarding the transaction for kWh between the third-party and the end-user. Additionally, as stated above, this diagram represents all the major functions as discrete even though there may be integration of some functions as the industry grows and matures.

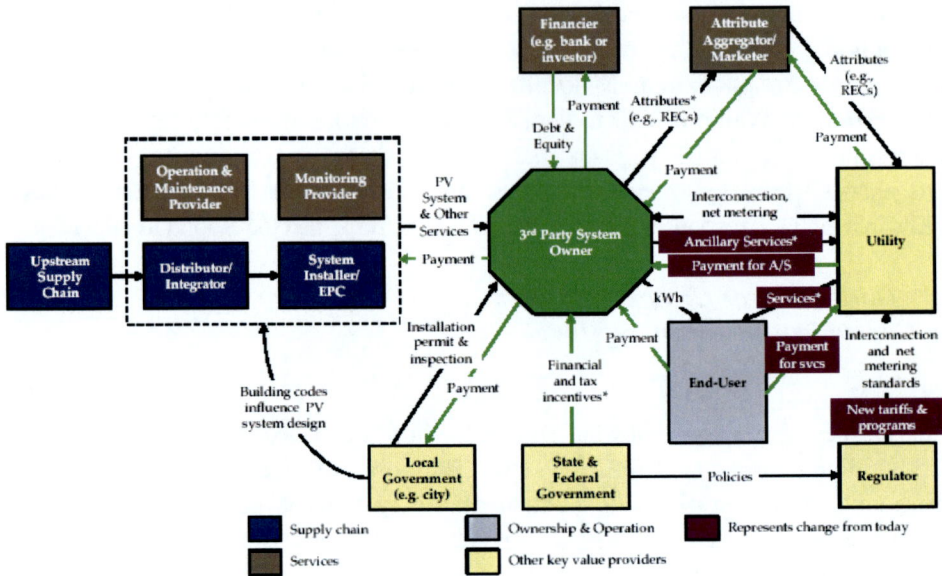

* Requires communications and control, including for performance based incentives.

Figure 4-16. Third-party/Customer Owned and Utility Controlled Value Network.

4.3.5.2.2. Business Model Requirements

The requirements for the utility controlled, third-party/customer owned business model are largely the same as for third-party/customer controlled model. The key difference is the regulatory regime, which would enable the utility to control significant assets on the customer side of meter. To the extent that utility control is not just for grid benefits but also to enable the utility to offer other services to the end-user, these regulatory changes will need to address the rules governing competition for providing these services. The main competitive issue is that the utility, as a monopoly, has an unfair advantage in its access to the customer. If the utility is allowed to access assets behind the meter for the benefit of the grid, but then is also allowed to leverage this access to offer customer-based services like backup power or energy management, other companies without such access might see this as unfair. To the extent that utilities were allowed to use the PV assets to provide value-added services to those customers who own them, the structure and pricing of these services must be determined in a transparent and equitable manner.

Since the asset is now controlled by a utility, the structure of the system-wide control architecture would be different. There is *less* of a need to be able to send the appropriate price or other signals to many owners. Instead, the control of the PV assets could be integrated into the utility's overall distribution network. For this, some key technologies include:

- Advanced communications, monitoring and metering, which allows for the utility to monitor, communicate with and control a large number of PV systems, and in a manner that best integrates with what the utility does on its side of the meter. Ideally, the network would be able to look at these PV assets in aggregate as opposed to discrete units;

- The ability to measure the value to the utility will be important because it will form the basis of the financial arrangement between the utility and the owner, and will affect tariff structures and cost levels;
- Energy storage to firm-up capacity and allow for dispatch of the system, and to enable services to end-users; and
- The ability to island the PV system, or the deployment of technologies that can lead to micro-grids, would enhance the value of the PV system to customers and the utility.

In these business models, the utility is more likely to provide incentives to site PV in constrained areas to maximize grid value. The utility also has more of an incentive to offer maintenance so that it can rely on the PV assets.

4.3.5.2.3. Variations
A number of variations on the basic business model described here are possible (see table 4-8).

Table 4-10. Business Model Variations for Third-party/Customer Owned and Utility Controlled PV

Business Model Variation	Description
Forward REC purchases	This is similar to the variation described above in table 4-7; however, the utility could offer financing to encourage PV deployment in critical areas, perhaps offering preferential terms to customers in areas where the PV would have the greatest value.
Customer-focus	This is similar to the variation describe above in table 4-7 except that PV could become part of a technology package offered to end-users by utilities, as opposed to a third party.
Distribution utility as network services provider	This model is similar to the one described above in table 4-7, except that the utility manages the asset on behalf of the customer/owner.

4.3.5.2.4. Challenges
Like the customer or third-party owned and controlled business models, the incremental value from active control of the PV assets may be too small to create new business models around these incremental value streams. Moreover, in these business models, the utility still does not control the timing or siting of PV capacity; therefore, it is difficult for the utility to incorporate deployment of PV into capital planning, as the ultimate decision to site remains out of their control. Nevertheless, as the utility gains experience in managing an ever-growing fleet of distributed PV, the utility should be able to gradually incorporate PV into its planning activities, which should increase its value. It may be important for the utility to enter into a long-term contract with the owner to enable it to include impacts of PV in system planning. As with the customer or third-party controlled business models, the utility could issue RFPs for customers to install PV and could include financing as part of the package. This would encourage siting in the best locations. The utility may also provide incentives to site in constrained areas to maximize grid value.

In unbundled markets, there is little way to address the potential impacts of widespread adoption of PV on retail electricity suppliers, wholesale generators, and transmission companies. However, these companies could take on the third-party ownership role, thereby mitigating some of the impacts. Still, in the case of very high PV deployment, there is potential for there to be stranded generation assets, in both vertically integrated and restructured markets.

In this structure, the dollars are expected to flow in both directions, and it is not entirely clear how this would work. The regulatory structure would need to make accommodations for lost revenues and the means to value PV attributes. Some activities may result in the flow of dollars to the utilities (e.g., for maintenance and power wheeling).

4.3.5.3. Utility Owned and Utility Controlled PV Business Model

4.3.5.3.1. General Description

This business model represents the greatest departure from today, as the utility reaches unequivocally behind the meter to own assets and provide a range of services to customers (see Figure 4-17).

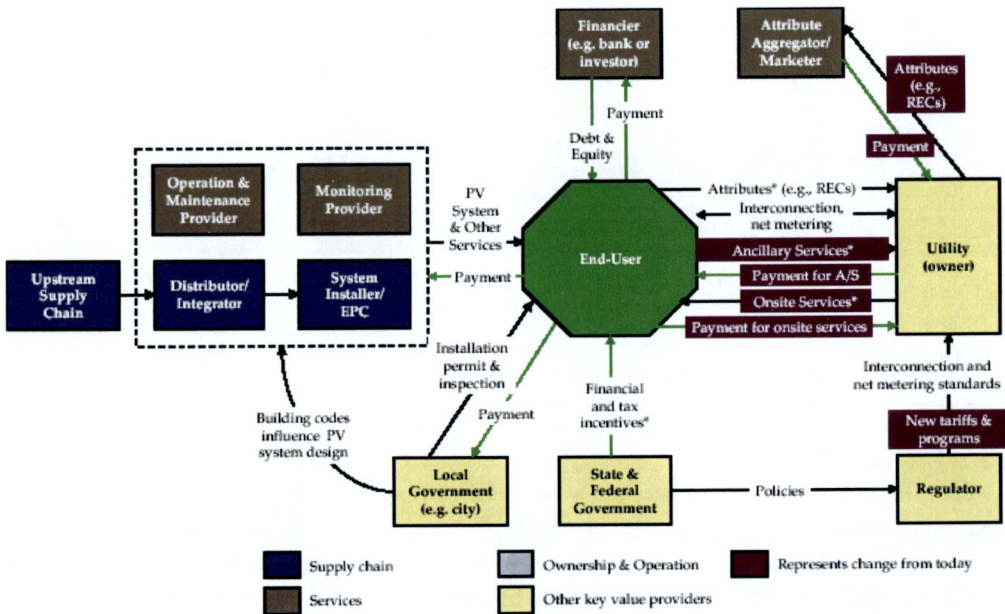

* Requires communications and control, including for performance - based incentives.

Figure 4-17. Utility Controlled and Owned Value Network.

This model seeks to unlock greater distributed PV value by involving the utility directly in both ownership and control of the asset, and in monetization of the asset's value. This arrangement fits well with utility core competencies of asset ownership and operation. Given that PV is a capital-intensive asset, there is merit in putting such utility-owned assets in the rate-base.

With utility ownership, the S-RECs would go to the utility; therefore, this model may work particularly well in markets with S-RECs where the utility is the obligated party. The utility could also bundle the S-RECs into a green pricing program or sell them to other parties. Since S-REC values can be high, this model could be an attractive means of compliance with solar set-asides.

By allowing the utility the greatest control over the placement and subsequent operation of the asset, this model should generate the greatest overall value for the utility. Moreover, in this model, the utility can readily incorporate the grid benefits into its basic cost of service, as well as sell value-added services to the end-user. Of the three groups of business models, this one is the easiest model for the utility to incorporate deployment of PV into their capital planning, as the ultimate decision to install is in their control. However, the issue of competition will be a complication as the utility could have unfair advantage in providing value-added customer-oriented (vs. grid oriented) services that a third party may want to provide.

Like the other business models described above, regulatory and policy regimes will need to be changed significantly to allow the utility to reach so overtly behind the meter. To mitigate the potential scope of such regulatory and policy changes, the PV systems could be located on customer premises but placed on the utility side of the meter (this is similar to SDG&E's placement of PV systems in its Sustainable Communities Program discussed in Section 4.2). In the past, states have prohibited utilities from owning and operating distributed energy resources (DER) because of concerns regarding market power. This concern will need to be addressed if and when PV systems become very inexpensive or otherwise attractive to utilities.

This business model will make sense in both vertically integrated and restructured markets, although the issue of generation asset ownership for distribution companies in restructured markets will need to be addressed.

This business model is expected to evolve more slowly than the others, given the additional regulatory changes required to permit utility control and ownership. Additionally, for utility control to have significant value to the utility, distributed PV has to exist on a sufficient scale, having a material impact on key values, such as the ability to offset generation, capacity, and T&D investments.

4.3.5.3.2. Business Model Requirements
The business model requirements are largely the same here as above. The key difference is the regulatory regime which would need to address issues of asset ownership on the customer side of the meter. Competition for the provision of customer-side services, which are not subject to state regulation, would need to be addressed if the PV assets are used to provide value-added services to those particular customers that use them.

In a business model in which the PV assets are both controlled and owned by the utility, the structure of the system-wide control architecture would be different than in models in which the customer or a third party either controls or owns the assets. There would be no need to be able to send other signals to a large number of owners. Instead, the control of the PV assets would be integrated into the utility's overall distribution network. Moreover, the deployment and use of PV systems would be more readily integrated into the utility's planning processes; PV systems would become extensions of the distribution grid. Thus, as PV is continually added, the utility would have the opportunity to make sure that the grid configuration remains optimal. This business model would also likely make it easier for utilities to justify investments required for grid reconfiguration, as this becomes necessary. For this model some key technologies include:

- Advanced communications, monitoring and metering, which allows for the utility to monitor, communicate with, and control a large number of PV systems, and in a manner that best integrates with what the utility does on its side of the meter. Ideally, the network would be able to look at these PV assets in aggregate as opposed to discrete units;
- The ability to measure the value to utilities will be perhaps less important than in other business models because the PV system is an extension of the distribution system and, therefore, simply part of the overall cost of service. Unless the utility offers products and services geared towards individual customers, such as backup, energy management, or different energy delivery pricing options, the arrangement between utility and the customer would be simple;
- Energy storage to firm up capacity and allow for dispatch of the system, and to enable services to end-users; and
- The ability to island the PV system, or the deployment of technologies that can lead to micro-grids, would enhance the value of the PV system to customers and the utility.

In these business models, the utility would identify the sites it wants to develop and work with those customers. If all that is needed is a site, a nominal fee could be charged for use of the space (like a lease for roof space). If value-added services are provided to the end user, a different arrangement will be required, based on the services provided. As the owner, the utility would almost surely provide the maintenance so that it can rely on the PV assets to the greatest extent possible.

4.3.5.3.3. Variations
A number of variations on the basic business model described here are possible (table 4-9).

Table 4-11. Business Model Variations for Utility Owned and Controlled PV

Business Model Variation	Description
Customer-focus	Similar to variation described above in table 4-8.
Distribution Utility as Network Services Provider	This model is similar to the one described above in table 4-8 except that the utility both owns and manages the asset on behalf of the customer/owner for the benefit of the system as a whole.
System Components	Utility owns the inverter because it is the interface point between utility and customer. It is the system component most likely to fail (the utility may have made investments in the grid to accommodate PV system), and it is the component that requires most on-going O&M. Utility influence on the inverter market could result in changes that improve integration with other distribution automation and smart grid activities.
Battery System	Utility offers to add a battery back-up system, including automatic islanding and resynchronization, to customer purchases of PV systems. They then charge the customer for additional reliability service. Incremental system cost would be rate based.

Business Model Variation	Description
Battery System and Controls	In addition to the battery back-up system, the utility adds controls to the system that allow it to tap into the system for additional value (e.g., peak dispatchability, VAR support).
Battery System, Controls and Plug- in Hybrid Vehicles	In addition to the battery system and controls above, the utility offers an interface for plug-in hybrid vehicles. By adding the plug-in hybrid vehicle load, the utility becomes a provider of transportation energy.
Transfer	Utility has agreement to purchase system from a third-party/customer once tax benefits and accelerated depreciation are utilized (typically 5-6 years). Utility is not the best party to act as a system developer or take advantage of tax credits and other benefits at this time. Utility is well suited to take on
	long-term O&M. System could be rate-based and would come at a significant discount to the utility, with much of the useful life of the system remaining. In addition, the utility could provide land or roof space to third- parties, thus reducing the cost of the system.
Flat Price Electricity	Utility interactively manages energy consumption with the customer. Alternatively, allow utility to monitor and control energy use so that the customer does not exceed certain price points. For example, a person on a fixed income would be provided with a mobile device that supplies information about the energy use. The goal is to not make it more complicated than it is today
Bundled Electricity Services	Manages electricity preference for customer, e.g., least cost, greenest. Incorporates demand response, energy efficiency, PV (possibly as "loss leader"), integration of plug-in hybrid electric vehicles. Takes advantage of fact that price of PV-generated electricity does not change (like long-term fuel contract)
Bundled Energy Services	Electricity (see above), but could also include provision of other energy services, including heating fuel. Could include partnership with biofuels and/or oil company

4.3.5.3.4. Challenges

Technological challenges would be not much different here than for other business models. The key challenges are likely to be regulatory, as these options require the greatest changes to the current regulatory structure. Regulations need to enable the utility to own customer-side assets, and, in the case of providing customer-oriented products and services, to fundamentally change the way utilities charge for service. The network may need to allow for distribution wheeling, which would entail significant reconfiguration. Note, however, that the issue of lost-revenues is more or less eliminated since the utility controls the output of the PV system and has the asset in the rate base.

In unbundled markets, there is little way to address the potential impacts of widespread adoption of PV on wholesale generators and transmission companies. Unlike in the previous business models, generation companies could not take on the third-party ownership role to help mitigate these impacts.

Tariff structures will need to incorporate rate-basing of PV assets into their structure and allow for the separation of charges for system-wide benefits and end-user benefits (where PV is sited and is part of service package). This could be a complex formula. In general, this option is tightly linked to how utility service offering will change in the future. PV could be the catalyst for this type of change.

5.0. Conclusions and Recommendations for Future Research

5.1. Conclusions

Currently, PV business models revolve around access to lower-cost financing, increasing the efficiency of the supply chain, and reducing hassles and complexity for the customer. These types of incremental improvements will occur naturally as *0* and *1st Generation* business models continue to evolve.

Up until this point, there has been little reason to address system control or consider PV aggregation as an explicit policy matter, given the limited number of PV systems installed on the distribution grid. However, a time will come—in some areas of the country much sooner than others—when the sheer number of installed distributed PV systems becomes a material and operational concern—or opportunity—for utilities. Policy and regulatory considerations will then be paramount.

The most significant finding in this study to date is that the full benefits of an extensive distributed PV resource are not likely to be realized without some degree of utility control and ownership. The need to have active management and control of an increasingly large number of distributed PV systems implies that utilities will most likely become more involved in one way or another. As market penetration increases, distributed generation will reach a scale (i.e., generally greater than 100 MW) that could translate to significant value. For example, utility involvement could help optimize distributed PV assets by incorporating them into grid and generation planning. This is likely to reduce new peaking power requirements, distribution substation upgrades, and other system investments, thus unlocking latent value in the electric grid as a whole.

The results of the analyses performed in this series of DOE studies show that the real value of PV lies in its potential to offset generation, capacity, and T&D investment. Such value greatly outweighs the value PV has for providing ancillary services on the distribution grid. Therefore, business model development will not be driven by the potential for ancillary grid services. It is the possibility that a large quantity of distributed PV systems will be installed that provides the greatest potential benefit to the nation's energy infrastructure, as these systems in aggregate could actually offset significant investment requirements in new generation, transmission, and distribution capacity.

Aside from the technological changes that will be required to accommodate a large capacity of PV on the grid, the organizational structure of today's utilities does not facilitate the adoption of the new business models discussed in this chapter. For example, current grid planning and operation practices do not explicitly take into account the potential value from PV, and these functions are largely separate within utility organizations, which hampers inclusion of PV and other distributed resources in system planning.

5.2. Recommendations for Future Research

5.2.1. Role for DOE

Through its efforts on Renewable Systems Interconnection, DOE is investing in understanding how technologies on the distribution grid can make significant contributions to meeting future electricity demands. Continued work on business models is a natural complement to this, as business models will facilitate how all of the technologies ultimately come together and transfer value to stakeholders. The future business models described in this chapter will require changes to industry structure, which implies risk for key stakeholders. DOE is in a position to work with key stakeholders to help mitigate some of the risk involved in pursing these new approaches.

To understand the potential real costs and benefits, promising future business models will need to be piloted at a sufficient scale, requiring significant time and investment. Today, the exact scope, duration, and scale of the business model pilots required is not clear because many issues still need to be addressed. It is critical that key stakeholders are engaged in understanding exactly what is holding back the development of these business models, because many of these companies and organizations are actively considering what the future will look like and how they will participate. In addition, explicit business model development should be coordinated with work on smart grid capabilities (e.g., distribution automation and advanced metering infrastructure), energy load management (e.g., demand response), and other distributed resource technologies. All of the business models discussed in this chapter will require integration with these other emerging technologies and capabilities, as well as PV. Because of the potential fundamental changes in regulation, technology, ownership, control, and grid management implied by the former, it is premature at this point for DOE to issue an RFP to pilot new business models. As an alternative to immediate pilot activity, we recommend the following three- phase approach which is illustrated in Figure 5-1:

- Phase 1: Build the foundation
- Phase 2: Develop the scope
- Phase 3: Pilot business models and fund other supporting activities

As described below, work in Phases 1 and 2 involves studies and preparation for future business model pilots. These first two phases will likely take one and a half to two years. This additional time will allow the industry to mature and the pace of PV deployment to increase such that it more likely to achieve a scale sufficient to support business model pilots in Phase 3.

5.2.2. Phase 1: Build the Foundation

In Phase 1, DOE would develop a better understanding of how future business models would work, the barriers they will face, and the steps that will be required to overcome these barriers. Phase 1 involves three types of activities: stakeholder engagement; collaboration with the smart grid, energy management, and non-PV distributed generation communities; and analytical studies. The activities proposed for this phase would likely cost $2-$4 million. Cost share would be minimal as this work would primarily be studies, as opposed to development and deployment of technology.

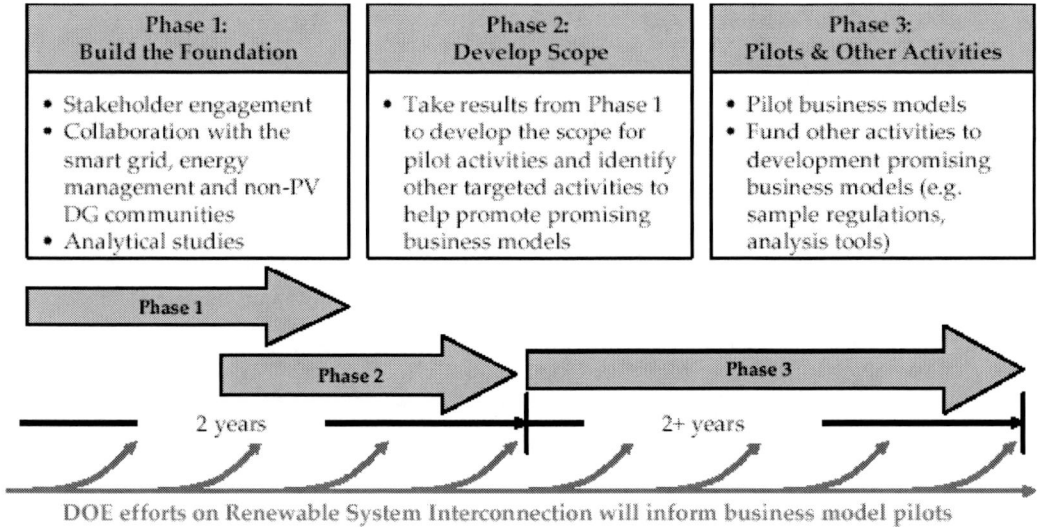

Figure 5-1. Three-phase approach to developing business model pilots.

5.2.2.1. Key Stakeholder Input

We suggest that DOE engage those stakeholders that are most critical to the success of these new business models. These are the companies likely to have the most significant financial stake in the outcome. The purpose would be to vet the business models developed in this paper and determine what is needed to move them forward. The key stakeholder groups we suggest engaging are utilities, regulators, and companies likely to provide equipment and services for these business models.

Utility leaders are already thinking about what the future will look like and what their role will be. This is especially true in states with aggressive climate change regulation in place. Strategic planners and other key decision makers from these utilities would likely have interest in participating in this type of activity with DOE.

Since it is largely the regulatory structure that defines utility business models, and new PV business models will require regulatory changes, it is critical to engage state and federal regulators to determine what types of actions they can take to facilitate promising business models. By increasing the understanding of the potential benefits business model options (like the ones presented in this chapter may have), the more likely it is that regulators will take stronger actions to support distributed generation, other technologies, and pricing structures that will be especially important in meeting key policy goals, such as climate change. Regulators may require guidance, tools, and information to better evaluate new PV business models. We suggest not just engaging regulators at the state level, but also engaging the National Association of Regulatory Utility Commissions (NARUC) and the Federal Energy Regulatory Commission (FERC), as both of these agencies have great influence over the decisions at the state-level. Regulators should be engaged in states that have demonstrated strong support for PV, and also in states with little regulatory support to date.

Finally, DOE should engage equipment and service providers who, like the utilities, are actively considering what the future distribution grid will look like and what product and services design features will be required to maximize value for their customers. Many of these companies have already developed future scenarios similar to the business models developed in this chapter, and their input on barriers and how DOE could help would be invaluable.

The type of questions these groups could address include:

- What are the key barriers restricting implementation of new business models?
- Do any of the new business models being considered have fatal flaws?
- How could the business models be modified or improved?
- What specific regulatory hurdles must be addresses?
- What could DOE or others do to encourage the most promising business models?

Various formats could be utilized for engaging stakeholders. We suggest that initially the different stakeholder groups meet separately so their specific concerns are thoroughly understood. Face-to-face meetings between a handful of high-level individuals would facilitate consensus building around priority issues. For regulators with limited budgets and time, DOE could consider a format that would not require additional expense for travel, such as a series of facilitated conference calls.

The current SEPA project, supported by DOE to engage utilities, might provide the answers to some of the questions listed above. In addition, DOE should review the results of the EPRI STAC work on utility incentives for distributed generation as it becomes available.[17]

5.2.2.2. Collaboration

It is recognized that there are several other technologies faced with similar issues as PV, and there is much that can be learned about business model development by working both internally and externally with DOE to leverage what has and is being learned by organizations working with these technologies. This includes technologies that may be integrated with PV systems (i.e., as battery storage) and technologies that may operate independently (i.e., fuel cells and combined heat and power (CHP)). There are also technologies such as Advanced Metering Infrastructure (AMI) and Smart Grid that are likely to prove necessary to facilitate the development of some of the business models that will be developed for PV. Many studies and programs have been developed to address similar issues for these other technologies and it will be important to leverage this work.

In preparation for developing the scope of the analytical studies and the subsequent pilot phase, we recommend that DOE develop an internal working group to establish internal collaboration between groups working on solar PV, smart grid, energy management, zero energy buildings, and non-PV distributed generation. Since the business models will require integrating PV with these emerging technologies and capabilities, collaboration across DOE's efforts will be critical. The focus of the collaboration would be to get input from these other groups on the business models developed in this paper and to identify opportunities for collaboration on the pilots

5.2.2.3. Analytical Studies

We recommend structuring several analytical studies that would serve to define more specifically the potential business and market structures to be tested during the business

---

[17] Both the SEPA and EPRI projects are described briefly in 2.0.

model pilots. For example, the business model structures outlined in this chapter will need to be developed in more detail to address how various regulatory structures will integrate with the business model(s) (e.g., municipal, competitive retail markets, vertically regulated utility markets) as cost, benefit, and value will flow to stakeholders differently depending on regulatory structure. For example, the types of questions that will need to be addressed include:

- How can distribution capacity value be determined for PV and allocated to the rate structure?
- Can distribution utilities obtain cost recovery for grid or customer sited PV?
- How will value be determined?
- How will generation capacity value be determined?
- How will generation capacity be allocated in the rate structure?

Another objective of the analytical studies is to develop a clear understanding of other key characteristics required to support successful implementation of each business model, including:

1   Technology bundles and performance characteristics (including the non-PV elements discussed above);
2   Utility organizational structure; and
3   Scale (e.g., amount of distributed generation required to make the model viable).

We suggest that DOE fund both generic and utility-specific studies. The generic studies would serve to advance understanding openly and could result in the development of tools for key stakeholder groups, such as utilities, regulators, policy makers, and equipment and service companies involved in providing energy services on the distribution grid. Generic studies could also be used to establish methodologies for evaluating the costs and benefits of business models such as the ones developed in this chapter. Finally, the results of generic studies could be used to raise awareness among key stakeholders.

In addition to the generic study, there is value in DOE funding utility-specific studies that would use information about specific system structures to evaluate business model costs and benefits. While DOE would have to determine to what level the specific utility information could be held confidential, these types of studies could have great value, especially to those utilities that have not begun to evaluate these types of business options.

Finally, both the generic and utility specific studies will benefit from the results of the Renewable System Interconnection studies and any additional DOE effort in this area. For example, additional research on the value of PV and likely characteristics of future PV systems will all be useful inputs to these studies.

We would expect the three activities in Phase 1 to result in:

- More robust descriptions of business model options;
- A specific list of prioritized items that would help promote promising business models; and
- Initial interest and input regarding the structure of a program to test pilot business models (feeds into Phase 2: Develop the scope).

5.2.3. Phase 2: Develop the Scope

In parallel with the activities identified in Phase 1, we recommend that DOE begin the process of working with the various stakeholders to develop the scope of desired pilot activities and identify other targeted activities that could help promote promising business models. Engagement of stakeholders in Phase 1 could be used as an opportunity to get input on desirable pilot activity and create interest. This work could be accomplished through a series of regional, topical, or stakeholder related DOE sponsored meetings and workshops, structured to solicit input from stakeholders on business model needs and barriers. Geographic regions could include the West Coast, Northeast, and Southeast. Topics may include things such as: "Barriers to Utility Ownership of PV" or "Creating Value for Utilities with Grid Sited PV."

At present, it is impossible to know what the most appropriate approach for piloting business models will be. While there is vast experience at DOE and elsewhere in piloting and demonstrating new technologies, government sponsored pilot programs to demonstrate business models are less common. In fact, it may be challenging to keep all stakeholders focused on the objective of piloting business models versus piloting or demonstrating technology.

Aspects of the pilot program scope that DOE should consider include:

- *Scale*: What scale is required to test the business model?
- *Geographic Location*: What locations will provide the best pilots and results? For example, areas where high penetration of PV already exists or where very little activity has occurred?
- *Cost share*: What will DOE want to cost share? At what level? Who are other likely funders?
- *Timing*: How will the need for regulatory approval impact the timing?
- *Participants*: Are most likely utility participants' distribution companies and load serving entities? What equipment and service companies might want to team with utilities?
- *Organization*: Given the potentially high cost of piloting business models, could utilities and companies form a consortium to test a business model in one service territory?

In addition to the pilots, DOE will want to consider funding other targeted activities that address the barriers and needs identified in Phase 1. These types of activities may include such things as:

- Model tariffs and standards
- Sample regulatory structures approved by NARUC
- Spreadsheet based tools for comparing and evaluating business model impacts

By working with the stakeholders early-on, DOE will be able to begin to understand the potential types of groups that are likely to propose business model pilots. Through a series of regional, topical, or stakeholder-specific workshops, DOE can begin to solicit interest in bidding and begin to engage in discussions with groups regarding the types of proposals they are likely to be able to assemble. Input gathered during these meetings and workshops would be used to structure the RFP in a way that would increase the likelihood of obtaining desirable proposals for Phase 3.

5.2.4. Phase 3: Pilot Business Models and Other Supporting Activities

The importance of running pilots is to demonstrate the value that can be created through new business models and show that they are possible. For most of the future business models described in this paper, significant changes to industry structure will be required, including changes to grid hardware, grid operation, utility organization, and utility regulation. To help prepare the industry and to smooth the road with regulators, demonstrations of the real costs and benefits are essential. In addition, pilots will serve to identify barriers and challenges that need to be addressed. This type of support from DOE could help avert a slow down in PV market expansion due to resistance from utilities and regulators to permit high penetration of PV on distribution systems.

As has been discussed in this chapter, development of effective PV business models will require other new technologies and systems. These may include battery storage, micro- grid system architectures, AMI, smart grid technologies, and integration with energy management systems. Business model pilots should be structured as cross cutting activities that benefit a broader set of technologies and stakeholders. Additionally, pilots should address a broad set of industry stakeholders; for example, municipal utility organizations, distribution companies, vertically integrated utilities, energy service companies, equipment providers, and regulators. As the pilot programs are structured, it will be important to assess how the various stakeholders and ancillary technologies are included.

Since DOE has made a great effort to understand what the future will look like and appears to be pushing forward on the technical aspects of the distribution grid of the future, it seems logical that DOE should help test the business models that are most likely to unlock the value technological advances can deliver.

REFERENCES

Dicum, Gregory. "Green Solar Gets Practical". 1/25/2006, © 2006 Hearst Communications, Inc. http://sfgate.com/cgi-bin/article.cgi?f=/g/a/2006/01/25/gree.DTL

FPL Group press release: FPL Group Plans to Boost U.S. Solar Energy Production. JUNO BEACH, Fla., Sep 26, 2007).

"Lennar to Include Solar Electric Systems in All New San Francisco Bay Area Homes", 03/01/2007, http://www.buildingonline.com/news/viewnews.pl?id=5849

Moore, Geoffrey. Crossing the Chasm, Harper Business, 1991. Navigant Consulting PV Service Program. August 2007.

Navigant Consulting, The Value of PV, developed as part of the Renewable Systems Integration Studies funded by DOE 2007.

Solar Shares presentation at Solar Power 2007 Conference Pre-Conference Workshop on Utility Business Models. Long Beach, California. September, 2007.

"Xcel Energy announces the largest photovoltaic central solar power plant in the United States". Xcel Energy News Release: Sept 25, 2006.

APPENDIX: CASE STUDIES OF CURRENT BUSINESS MODELS

7.1. End-user Owner/Residential Retrofit: Borrego Solar

Borrego Solar Systems, Inc. has more than 25 years of experience in the solar installation industry, specializing in commercial, residential, and public-sector turnkey, grid- connected systems. The case study (table 7-1) focuses specifically on Borrego's end-user, residential retrofit business model, which fits into the category of Hassle-Free, as described in table 4-2.

Table 7-1. Case Study: Borrego Solar

Owner/Application	User/Residential Retrofit
Company Profile	
Business Model Variation	• Hassle-free
Company	• Borrego Solar Systems, Inc.
Consumer Profile	• Demographic focus: Borrego's customers tend to be: homeowners/professionals, upper-middle class, living in 1,500-2,000 sq-ft homes, and have $100-$150/month electric bills. [18] • Geographic focus: The San Francisco Bay area and Southern California, with a recent push into the NE region of the United States. • Typical system size: 5 kWp • Total system cost: Ranges from $25K to $1 00K, or $8.00 -$1 0.00/Wp. • Net cost: Borrego cites an example of a $30,400 installation translating to a net cost to owner of approximately $19,000 after all possible rebates.
System Characteristics	• Economic breakeven point: 6-12 years • System technology: In its new market in the Northeast, Borrego is partnering with Evergreen Solar, the manufacturer that uses String Ribbon wafer technology. However, the company relies primarily on crystalline modules, along with grid-connected inverters from SMA and Xantrex. The company uses Web-enabled monitoring from SMA; it does not do battery back-up systems.
Marketing Process	• Reputation: Borrego has 27 years of experience in the California market, and one of the strongest reputations for turnkey solar systems. • Partnerships: Borrego builds relationships with home builders, solar suppliers, and government to develop new business.
Sources of Financing	• Cost reduction from incentives: State rebates cover 20-30% of total installation cost ($2.20/Wpac), along with $2000 in federal tax incentives. • Other: Borrego encourages customers to make use of secured loans, such as home equity loans (or lines of credit), which typically have the best terms and lowest interest rates.
Main Sources of Value Provided	• Processing and management of the entire process: Borrego creates a packaged deal and quotes the total cost to the customer at a post-rebate amount.

---

[18] Dicum, Gregory. "Green Solar Gets Practical". 1/25/2006, © 2006 Hearst Communications, Inc. http://sfgate.com/cgi-bin/article.cgi?f=/g/a/2006/01/25/gree.DTL

Table 7-1. (Continued).

Owner/Application	User/Residential Retrofit
Perspectives on the Market	
Core Competencies of Company	• Hassle-free process: Borrego streamlines the PV system design and installation process and quotes the system at a post-rebate price. • Brand image: Borrego has deep market penetration in California. Its strong brand image is based on quality and cost-effectiveness, and customer satisfaction/referrals. • Warranty offer: Borrego offers a 10 year warranty, including a payment for lost energy in case of system failure.
Value Provided to Customer	• Hassle-free process: Borrego uses a turnkey approach. It takes care of the entire PV system installation process, including the application for state/federal rebates. • Customer service: Borrego continues to monitor its existing installations to improve service and quality.
Key Challenges	• Cost: Initial investment in a PV system is still a deterrent to a customer. Borrego, like all suppliers, must try to offer cheaper systems as competition increases.
Future Innovations	• Third-party ownership: Borrego has partnered with Sun Run, a distributed renewable power company, to create a pilot program for Contra Costa County homeowners. Sun Run owns and operates residential solar systems and sells the electricity at a discounted rate. Thus, the residential end-user does not incur the up-front investment cost of the PV system.

Though the end-user owner/residential retrofit market is now the slowest growing of the U.S. grid-tied market segments examined in this chapter, Borrego's new third-party ownership innovation with Sun Run, as described in table 7-1 above, opens a door for potential customers who are interested in solar but do not want, or cannot afford, to incur the installation cost. Borrego has set the stage for the future development of third-party ownership and operation in the end-user owner/residential retrofit market.

7.2. End-user Owner/Residential New Construction: Old Country Roofing

Old Country Roofing (OCR) is a market leader in the end-user owner, residential new construction market. It was chosen as a case study because few similar business models currently exist, and competition in this market is expected to grow as builders and homeowners begin to understand the delicate relationship between roof integrity and solar roof installations. Table 7-2 highlights the key takeaways of OCR's business model.

OCR has recognized how little developers and builders know about the impact of solar installations on rooftop integrity. While the company is holding workshops to educate its customers in California, solar roofing is taking off across the country, and OCR recommends that the government take a role in supporting the education of solar roofing. OCR anticipated a potential new business service in 5-10 years, when existing homes with solar installations begin to have structural problems with their roofs.

Table 7-2. Case Study: Old Country Roofing

Owner/Application	User/Residential Retrofit
Company Profile	
Business Model Variation	• Roofing company
Company	• Old Country Roofing (OCR)
Consumer Profile	• Demographic focus: OCR serves early adopters in this new solar roofing market. Home builders are increasingly incorporating OCR's solar roofing into multi-unit community-home projects. • Geographic focus: Primarily Central and Northern California, but OCR recently opened an office in Nevada.
System Characteristics	• Typical system size: 1.5 - 3.0 kWdc; average is approximately 2.4 - 2.5 kWdc • Total system cost: A PV roof is 4-5 times the cost of a regular roof: $8-9/kWdc or $20-25k for the PV and $6-8k for the normal roof. • Economic rational: OCR tells customers to analyze breakeven from a cash flow or ROI point of view (i.e., look at incremental cash cost for a solar roof built into the mortgage price vs. the incremental cash savings on energy). With the PV cost built into a mortgage, a payback can be seen on the first utility bill. For example, if the additional cost of the PV system is $110/month, but the PV system is designed to lower the electric bill $120/month, that means "the house is paying the owner $10/month." • System technology: OCR's primary supply partner is BP. It frequently uses BP's Solar EnergyTile™ roof integrated solar modules and BP Solar Integra® low profile solar modules.
Marketing Process	• Three primary initiatives: 1) Sales teams for existing customer relations with homebuilders; 2) Participation in San Francisco PCBC trade show, which serves residential builders and their project teams and has up to 25,000 attendees; and 3) Free workshops to educate builders on solar roofing.
Sources of Financing	• Cost reduction from incentives: $2.60/Wpac from California New Home Solar program and a $2000 federal tax credit for residential systems. • Home mortgage: Homeowners roll the installation cost into their home mortgage.
Main Sources of Value Provided	o System installation: OCR was the first contractor in the country to offer a comprehensive turnkey solution that includes both roofing and solar.
Perspectives on the Market	
Core Competencies of Company	o Knowledge of roofing: OCR takes responsibility for the entire roof, not just the solar panel installation. They understand the importance of integrating the two to ensure roof integrity. o Economies of scale: OCR is the leading roofing company in northern CA; other roofing companies have not been able to scale. o Brand image: OCR is the first roofing contractor in the country to be certified "National Housing Quality Certified Trade Contractor" by the National Association of Home Builders.
Value Provided to Customer	• Hassle-free process: OCR provides an entire-package: design, installation, warranty, and customer service. • Customer service: OCR guarantees customer satisfaction, which was a "pinnacle decision maker" for Tim Lewis Communities in Sacramento.[19]

---

[19] Sacramento homebuilder, Tim Lewis Communities, announced plans to build several of its new communities to exceed state standards for energy efficiency, certifying them as a California Green Builder. Three communities will offer Solar Living Homes and were available for pre-sale beginning in late March and mid-April 2007. http://www.sustainablehomemag.com/CDA/Articles/Feature_Article/BNP_GUID_9- 5-2006_A_1 0000000000000 144641

Table 7-2. (Continued).

Owner/Application	User/Residential Retrofit
Key Challenges	• Consumer awareness: Builders do not fully understand the solar roofing concept. They are still uneducated and may not choose OCR over a stand-alone solar installation company because they do not understand the value of roof integration. • Investment rational: Most customers want to know how many years it will take to recoup their investment. OCR is shifting the focus to a cash flow analysis to show that a PV system generate positive cash flows starting in the first month.

7.3. Third-party Owner/Commercial Retrofit: SunPower

SunPower has more than 10 years of experience in providing roof, ground, and elevated-parking solar systems to commercial customers. The company most recently signed a 10- year PPA-to-Own with Macy's Department Stores to provide a hassle-free energy solution to 15 of its stores in California. Table 7-3 provides the details of SunPower's Basic/PPA-to-Own business model.

Table 7-3. Case Study: Sun Power

Owner/Application	Third-party Owner/Commercial Retrofit
Company Profile	
Business Model	• PPA-to-Own
Company	• SunPower Corporation Systems
Consumer Profile	• Demographic focus: Clients include over 15 Fortune 500 global corporations. Positive public relations and cost reduction motivate these customers. • Geographic focus: SunPower Systems has a global focus with revenue divided between its U.S., Europe, and Asia operations.
System Characteristics	• Typical system size: 1 MW • Total system cost: System Cost is not published by SunPower as it can vary by location, solar irradiance, the purchase type, and many other factors. • Payback period: 5 to 15 years
Marketing Process	• Focus: Large accounts through direct marketing via sales representatives. • Modes of outreach: A variety of media and outlets are used, including: print and web collateral, active public and media relations, print and radio advertising, industry and customer alliances, and participation in renewable energy industry and regional trade conferences and events.
Sources of Financing	• PPA: The customer signs a 10-year (minimum) fixed price agreement with Sun Power. The customer does not own the PV system, but simply purchases the energy it generates. • Investment partners: *GE Finance, Morgan Stanley, MMA Renewable Ventures, others.*
Owner/Application	Third-party Owner/Commercial Retrofit
Main Sources of Value Provided	o Processing and management of entire process: The customer simply buys the solar kWh produced at a fixed-rate for at least 10 years. SunPower pays for and deploys the solar system, and the customer hosts the system.
Perspectives on the Market	
Core Competencies of Company	o Technology: Best-in-class non-penetrating rooftop mounting and single-axis tracking systems are key competitive advantages. o Project support: R&D programs and large-scale project experience.

Value Provided to Customer	• One-stop shop: Sun Power has access for those customers that want a PV system without the responsibility. Commercial customers are able to simply buy the solar power a PV system generates.
Key Challenges	• Extending and enhancing U.S. federal and state incentives for commercial solar power purchases. • Reducing the cost of solar power systems by 50% by 2012 in order to compete with retail electric rates. • Generating a broader awareness for the benefits of solar.
Future innovations	• System cost reduction: The supply of solar cells and components for panels and systems will rise and new technology will improve the way in which solar is delivered and installed, all contributing to a decrease in the cost of solar systems.

SunPower and its competitors are finding success with this business model primarily in California and the Northeast, where state financial incentives are the highest. If other states follow suit, the cost of electricity versus solar rises, or solar system costs come down, big box stores throughout the remainder of the country may become future customers.

7.4. Third-party owner/Grid-Sited: SunEdison

Table 7-4. Case Study: SunEdison

Owner/Application	Third-party Owner/Grid-sited
Company Profile	
Business Model Variation	• Basic
Company	• SunEdison
Consumer Profile	o Demographic focus: Typical customers of grid-sited applications (e.g., utilities) have the following characteristics: they are subject to renewable portfolio standard targets; have transmission and distribution systems that are facing specific areas with peak load congestion; are under increasing pressure from the public and customers to adopt renewable energy solutions; and have an energy generation portfolio with high exposure to fossil fuel volatility. • Geographic focus: SunEdison has a presence through the United States and Canada. The majority of its business comes from CA, NJ and CT, however, the company's business is growing in CO, AZ, NV, HI and RI.
System Characteristics	o Typical system size: >1MWp o Total system price: Averages $6.20/Wp dc • System technology: Sun Edison is one of the few solar services providers that are technology agnostic. Most solar service providers in the utility space have "proprietary technology" from trackers, to concentrators, to low-efficiency thin-film technologies. Sun Edison's innovation is around bringing in-house construction capabilities and fully optimized financing to reducing the non-technology costs of the projects. Matched with its technology approach, Sun Edison serves as a partner to the utility
Marketing Process	• Focus: On potential utility accounts through direct marketing via sales representatives.

Table 7-4. (Continued).

Owner/Application	Third-party Owner/Grid-sited
Sources of Financing	• PPA: A 20-year, fixed price agreement is settled between SunEdison and the customer. The rate is dependent on the available subsidies. The Alamosa project came in around 20 cents per kWh. • Investment partners: Goldman Sachs, MissionPoint Capital Partners, Allco, and one individual investor. • System cost reduction based on incentives: 30% Investment Tax Credit, 5-year accelerated depreciation of equipment, and any applicable state rebate.
Main Sources of Value Provided	o Processing and management of entire process: SunEdison handles everything from conducting the on-site assessment of needs and solar potential to selecting the appropriate technology and maintaining the finished PV system over its lifetime. SunEdison runs and manages their own trained and qualified crews, enabling them to maintain quality and accountability.
Perspectives on the Market	
Core Competencies of Company	o Access to a wide range of technology choices and commercial real estate. o Financial engineering skills (access to low cost of capital). o A strong track record.
Value Provided to Customer	• One-stop shop: No capital investment is required for the solar system host facility. SunEdison secures financing and takes responsibility for the entire PV system. The customer gets the benefit of a secure fixed energy rate.
Key Challenges	• Lowering inherent business risk given policy that could significantly increase or decrease the cost of solar to the customer. • Evaluating the many new technologies coming on the market each year to meet the needs of our customers. • Cost effectively maintaining the solar assets to achieve 99%+ system availability.
Future innovations	• New technologies: SunEdison may expand its model to long-term energy efficiency and DSM technologies with 7+ year paybacks, but currently it is focused on solar products. • International growth: SunEdison recently appointed a new executive to oversee SunEdison's solar deployment strategies in markets outside of the United States.

In Mid 2006, SunEdison and Xcel Energy signed an agreement for an 8 MW PV plant in Colorado in response to the state's new RPS requirement. The plant will use both flat-plate solar panels and concentrating photovoltaic units, representing 6.8 MW and 1.2 MW, respectively.[20] This system was completed in November 2007 and now represents the largest utility scale Solar PV project in the US. Xcel Energy is one of a hand-full of utilities across the US that has taken the initiative to pursue this sort of project to meets its RPS but is hopefully setting the standard for others to follow.

---

[20] "Xcel Energy announces the largest photovoltaic central solar power plant in the United States." Xcel Energy News Release: Sept 25, 2006.

In: Renewable Energy Grid Integration
Editor: Marco H. Balderas

ISBN: 978-1-60741-324-0
© 2009 Nova Science Publishers, Inc.

Chapter 4

PHOTOVOLTAICS VALUE ANALYSIS[*]

J.L. Contreras, L. Frantzis, S. Blazewicz, D. Pinault and H. Sawyer

PREFACE

Now is the time to plan for the integration of significant quantities of distributed renewable energy into the electricity grid. Concerns about climate change, the adoption of state-level renewable portfolio standards and incentives, and accelerated cost reductions are driving steep growth in U.S. renewable energy technologies. The number of distributed solar photovoltaic (PV) installations, in particular, is growing rapidly. As distributed PV and other renewable energy technologies mature, they can provide a significant share of our nation's electricity demand. However, as their market share grows, concerns about potential impacts on the stability and operation of the electricity grid may create barriers to their future expansion.

To facilitate more extensive adoption of renewable distributed electric generation, the U.S. Department of Energy launched the Renewable Systems Interconnection (RSI) study during the spring of 2007. This study addresses the technical and analytical challenges that must be addressed to enable high penetration levels of distributed renewable energy technologies. Because integration-related issues at the distribution system are likely to emerge first for PV technology, the RSI study focuses on this area. A key goal of the RSI study is to identify the research and development needed to build the foundation for a high-penetration renewable energy future while enhancing the operation of the electricity grid.

The RSI study consists of 15 reports that address a variety of issues related to distributed systems technology development; advanced distribution systems integration; system-level tests and demonstrations; technical and market analysis; resource assessment; and codes, standards, and regulatory implementation. The RSI reports are:

---

[*] Excerpted from *Subcontract Report* NREL/SR-581-42303, dated February 2008.

- Renewable Systems Interconnection: Executive Summary
- Distributed Photovoltaic Systems Design and Technology Requirements
- Advanced Grid Planning and Operation
- Utility Models, Analysis, and Simulation Tools
- Cyber Security Analysis
- Power System Planning: Emerging Practices Suitable for Evaluating the Impact of High-Penetration Photovoltaics
- Distribution System Voltage Performance Analysis for High-Penetration Photovoltaics
- Enhanced Reliability of Photovoltaic Systems with Energy Storage and Controls
- Transmission System Performance Analysis for High-Penetration Photovoltaics
- Solar Resource Assessment
- Test and Demonstration Program Definition
- Photovoltaics Value Analysis
- Photovoltaics Business Models
- Production Cost Modeling for High Levels of Photovoltaic Penetration • Rooftop Photovoltaics Market Penetration Scenarios.

Addressing grid-integration issues is a necessary prerequisite for the long-term viability of the distributed renewable energy industry, in general, and the distributed PV industry, in particular. The RSI study is one step on this path. The Department of Energy is also working with stakeholders to develop a research and development plan aimed at making this vision a reality.

ACKNOWLEDGMENTS

We would like to acknowledge the input provided by industry experts such as Christy Herig, Thomas Hoff, Ed Kern, Peter Kobos, Ben Kroposki, Robert Margolis, Richard Perez, Howard Wenger, and EPRI.

LIST OF ABBREVIATIONS AND ACRONYMS

AE	Austin Energy
CAISO	California Independent System Operator
CAPM	Capital Asset Pricing Model
CEC	California Energy Commission
CO2	Carbon Dioxide
CSE	Centre for Sustainable Energy
¢/kWh	Cents per Kilowatt Hour
DG	Distributed Generation
DOE	Department of Energy
E3	Energy and Environmental Economics, Inc.
EC	European Commission
ELCC	Effective Load Carrying Capacity factor

ERC   Emission Reduction Credits
GHG   Green House Gas
GT   Gas Turbine
MTC   Massachusetts Technology Collaborative
MW   Megawatt
NCI   Navigant Consulting, Inc.
NG   Natural Gas
NOx   Nitrogen Oxide
NPV   Net Present Value
NREL   National Renewable Energy Laboratory
NYMEX   New York Mercantile Exchange
O&M   Operation and Maintenance
PG&E   Pacific Gas & Electric
PV   Photovoltaics
REC   Renewable Energy Credits
R&D   Research and Development
SOx   Sulfur Oxide

EXECUTIVE SUMMARY

This chapter is part of a set of studies launched by the U.S. Department of Energy (DOE) and the National Renewable Energy Laboratory (NREL) to define a research agenda that will advance and enable a high penetration of renewable energy into the existing electricity grid. It specifically examines the value of photovoltaic (PV) systems to participating customers, utilities/ratepayers, and society. The study reviews existing published reports on the value of PV, summarizes the methodologies and quantification of PV values, and identifies research and development (R&D) that needs to be completed to fill in knowledge gaps.

PV Values

We identified 19 key values of distributed PV. These values are described in table E- 1.

Quantification of PV Values

PV values were quantified and allocated to several categories of stakeholders: customer participant, utilities/ratepayers, and society. Customer participants take the perspective of PV system owners and end-users. Utilities/ratepayers represent all electric utility customers in the region. Finally, society represents the general population.

Table E-1. PV Values

PV Value	Description
Central Power Generation Cost	PV systems produce electricity, reducing the amount of electricity that needs to be generated at other plants, which in turn decreases fuel and other O&M costs.
Central Power Capacity Cost	PV indirectly avoids and/or defers central power plant capacity investments by reducing demand-side consumption. Generation capacity value is the economic value of the avoided and/or deferred incremental resource (typically natural gas turbine) reflecting PV's peak load reduction. NIMBY opposition and higher construction costs are driving capacity costs up. PV also avoids the cost of running more expensive plants during peak loads.
T&D Cost	PV avoids and/or defers transmission and distribution capacity investments by reducing demand-side consumption. Transmission and distribution capacity value is the economic value of the avoided and/or deferred incremental resource reflecting PV's peak load reduction. NIMBY opposition and higher construction costs are driving capacity costs up. This value is also applicable to situations involving significant congestion issues.
System Losses	Avoided electric system losses are an indirect benefit because they increase the value of other benefits including energy production, generation capacity, environmental and T&D capacity.
Ancillary Services	Utilities can use inverters in PV systems to provide reactive power back to the grid. This increases power quality and could avoid the installation of capacitors.
System Resiliency	Significant deployment of PV systems coupled with storage could provide disaster recovery benefits.
Hedge Value	Current electricity generation is heavily dependant on natural gas and coal. Recent environmental constraints suggest that utilities will become more dependent on natural gas. PV lessens the exposure of the utility to volatile fuel prices and provides stable and predictable electricity prices.
Market Price Impacts/Elasticity	The elasticity of demand for electricity supply increases with more PV. Increased demand for PV may decrease the price of electricity from PV, spur market development, thereby further reducing the cost of PV. A decrease in the cost may then increase the demand for this lower cost good.
Customer Electricity Price Protection	Since there is no fuel expense, the costs of electricity from PV will not increase over the life of the system due to fuel costs and the consumer effectively locks in an electricity price.
Customer Reliability	PV can provide electricity to the PV owner during outages because it is not dependent on the grid. The electricity during an outage is limited to sunlight availability. Storage systems could help offset the intermittency issue and increase the reliability value to the owner.
Criteria Pollutant Emissions	PV systems eliminate criteria pollutant emissions (e.g., NO_x, SO_x) associated with non-renewable generation resources. Health benefits associated with reduced emissions are included in this value.
Greenhouse Gas Emissions	PV systems eliminate greenhouse gases (CO_2) associated with non-renewable generation resources. Renewable Portfolio Standards (RPS) and Renewable Energy Certificates (REC's) are common mechanisms to value emission reductions from renewable sources of energy such as PV.
Implicit Value of PV	The intrinsic societal value of PV to customers (e.g., environmental friendliness, feeling good, early adopter) and utilities (e.g., public relations, regulator compliance).
PV System Equipment and Installation Cost	The cost of equipment (PV array, inverter, battery, transformer etc.), engineering design and construction for a PV system.
PV System O&M Cost	This accounts for operations and maintenance expenses for the PV system. As PV cost is almost entirely capital, ongoing O&M costs are negligible.
Benefits Overhead	Costs associated with capturing and monetizing the value streams. Includes program administration, marketing and equipment costs (e.g. advanced metering).
PV Owner Electricity Bill	PV systems produce electricity therefore owners consume less electricity from the electric power system reducing their electric bill and the revenues to utilities/ratepayers.
Federal Incentives	Federal rebates offered to promote residential and commercial PV system that offsets initial investment costs though federal tax credits.
State Incentives	Incentives offered by States to promote PV systems, either through initial investment rebates or production credits.

We analyzed existing methodologies to quantify the PV values. A range for each value was calculated, because PV values have multiple drivers. Table E-2 summarizes the PV value ranges. On average, the values with the highest net benefits are central power generation cost savings, central power capacity costs, transmission and design (T&D) costs, greenhouse gas

(GHG) emissions, criteria pollutant emissions, and implicit value of PV. The value with the highest net cost is PV equipment & installation.

Table E-2. PV Values Ranges

	PV Values	Customer/ Participant (¢/kWh)	Utility/ Ratepayers (¢/kWh)	Society (¢/kWh)	Net (¢/kWh)	Value Drivers
Benefits	Central Power Generation Cost		3.2 – 9.7		3.2 – 9.7	Gas price, heat rate
	Central Power Capacity Cost		1.1 – 10.8		1.1 – 10.8	Effective load carrying capacity factor, gas turbine capital cost, life adjustment
	T&D Costs		0.1 – 10.0		0.1 – 10.0	Location, growth, climate
	System Losses		0.5 – 4.3		0.5 – 4.3	Location, time period, other benefits
	Ancillary Services		0 – 1.5		0 – 1.5	Ancillary service prices, voltage support
	System Resiliency		Low		Low	Quantification methodology unclear
	Hedge Value		0 – 0.9		0 – 0.9	Gas price forecasts, futures, heat rate
	Market Price Impacts/Elasticity		Low		Low	Quantification methodology unclear
	Customer Price Protection	0.5 – 1.0			0.5 – 1.0	Calculation method
	Customer Reliability	Low			Low	Quantification methodology unclear
	Criteria Pollutant Emissions			0.02 – 2.0	0.02 – 2.0	Market value of emissions
	Greenhouse Gas Emissions			0.02 – 4.2	0.02 – 4.2	Reduction costs, market value, discount rate
	Implicit Value of PV			0 – 2.0	0 – 2.0	Customer willingness to pay premium
Costs	Equipment and Installation	(47) – (19)			(47) – (19)	Size, location
	PV O&M Expenses	(0.15) – (0.05)			(0.15) – (0.05)	Type of system
	Benefits Overhead			(0.2) – (0.1)	(0.2) – (0.1)	Infrastructure and administrative costs
Transfer	PV Owner Electricity Bill	1.1 – 33.0	(33.0) – (1.1)		-	Customer type, rate structure, load profile
	Federal Incentives	1.58 – 7.95		(7.95) – (1.58)	-	Customer type, size, cap
	State Incentives	0 – 17.8		(17.8) – 0	-	State, customer type, size, production, cap
	Stakeholder Total	(43.97) – 40.7	(28.3) – 36.0	(25.7) – 6.6	(41.9) – 27.3	

Drivers of PV Value

The two main drivers for the highest magnitude values are location of the PV system and output profile or timing of the power output of the system. As illustrated in Figure E-1., a PV system will have higher benefits when it is located in a highly congested distribution system, where there is high insolation to increase production of the PV system, and where gas prices are high. PV systems will also have higher net benefits when a large share of their production is during peak demand periods, when the systems can displace expensive peaking plants, which have lower efficiency and utilization, and use more expensive fuel.

Central Power Generation Cost
Natural gas-fueled power plants are used as the marginal generation resource in many regions of the United States. As a result, natural gas prices in the region and the marginal resource heat rate (i.e., the amount of gas consumed to generate a kilowatt-hour [kWh]) are two key drivers of this value. Given these drivers, a PV system in a region with high gas prices that generates most of its power during peak-time and displaces electricity from a peaking power plant with a high heat rate will have a higher benefit than a PV system in a region with low gas prices that produces most of its power off-peak, displacing electricity from a baseload power plant with a low heat rate. A strategy to improve the benefit from this

value is to increase production from the PV system by optimizing orientation (i.e., latitude and tilt) and using tracking systems.

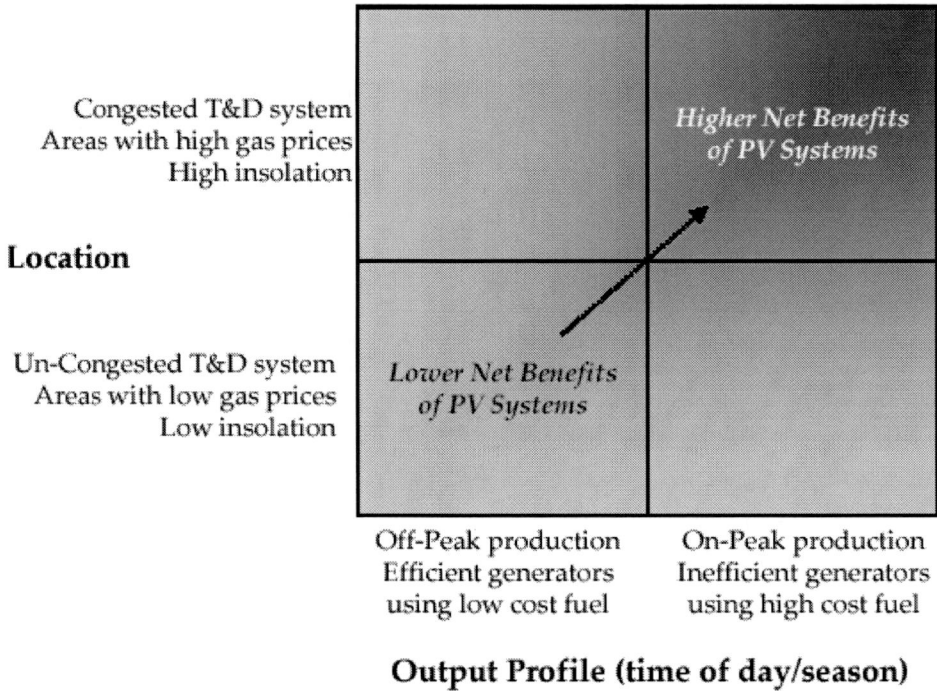

Figure E-1. Key Drivers of PV Values.

Central Power Generation Cost

Natural gas-fueled power plants are used as the marginal generation resource in many regions of the United States. As a result, natural gas prices in the region and the marginal resource heat rate (i.e., the amount of gas consumed to generate a kilowatt-hour [kWh]) are two key drivers of this value. Given these drivers, a PV system in a region with high gas prices that generates most of its power during peak-time and displaces electricity from a peaking power plant with a high heat rate will have a higher benefit than a PV system in a region with low gas prices that produces most of its power off-peak, displacing electricity from a baseload power plant with a low heat rate. A strategy to improve the benefit from this value is to increase production from the PV system by optimizing orientation (i.e., latitude and tilt) and using tracking systems.

Central Power Capacity Cost

The key driver for this value is the coincidence of peak demand with system output. Another key driver is the type of generation asset displaced. Peaking plants typically have lower capital costs than baseload plants. However, a peaking plant that runs a limited number of hours per year will have a higher capital cost per kilowatt-hour than a baseload plant. Given these drivers, a system that produces a high share of its output during on-peak hours and displaces a peaking plant will have a higher benefit. Various strategies to increase production during peak demand periods and increase the benefit from this value include: integrating energy storage to the PV system, and integrating load management applications with the PV system controls.

T&D Cost

While this value has significant potential, it has been difficult to capture. This value depends on the location of the PV system as well as the output during the T&D system's peak period. Locations with congested transmission and/or distribution systems typically require expensive upgrades that could be deferred where PV systems are installed to reduce congestion. Although this value includes both transmission and distribution, there are cases that are specific to one or the other. For example, PV can be installed in an area that reduces the transmission peak, but is in a distribution area with excess capacity, providing limited value to the distribution system. However, some of the most congested areas have network distribution systems in which interconnection standards currently prohibit or severely limit interconnections. And in non-network distribution systems, the deferral depends on the production of the PV system during the peak of the specific distribution area, which varies across the distribution system and can be a different peak period than the regional generation peak.

Many distribution planners will also want "physical assurance" (i.e., guarantees that the load the PV is serving is permanently displaced). Another barrier for this value is the potential for circuit overload following an outage or a recloser operation. Current interconnection standards (e.g., IEEE 1547) prohibit PV systems from riding through outages, leaving the T&D system to support the loads that would otherwise be served by the PV systems. For a local utility to defer T&D upgrades and capture the benefits of this value, it will need to assure that the PV system will effectively eliminate a certain load during its peak congestion period and that it will not suffer a load surge after outages. PV inverter-based load management systems would allow PV systems to capture the T&D cost benefits. Other strategies to increase the benefits captured from this value are to improve the ability to install PV systems in congested areas with limited roof space, and to firm PV output with storage and/or demand response.

Greenhouse Gas

This value is driven by two key factors: the amount of emissions displaced by the PV system and the value of the displaced emissions. PV systems have no point source emissions at the demand site, and therefore displace all the emissions otherwise associated with siting the marginal central generation resource. In most cases, the marginal resource will be a gas-fueled central generation plant. The higher the heat rate, the higher the displaced emissions. Coincidence of peak demand with PV system output also plays a role in this value as peaking plants tend to have a higher heat rate than baseload plants, producing higher emissions. There are several ways to value displacedGHG emissions: placing a price on carbon through a carbon tax, employing a cap-and-trade program, or creating renewable portfolio standard/renewable energy credit markets. The variety of valuation mechanisms has created a wide range of economic benefits for this value. Moreover, the climate and energy context is evolving quickly and producing an upward trend in the future economic value of the emissions reduction. A strategy to increase the benefits from this value is to support the development and adoption of a uniform valuation standard across the country. A secondary strategy is to increase the amount of displaced emissions by aligning the PV system production with peak demand periods, when dirtier peaking generation plants are the marginal resource.

Criteria Pollutant Emissions

Two methods of determining the value of offsetting criteria pollutant emissions are the avoided penalty/cost and the health benefits. These methods involve assessing the public health impact, regional air quality district emissions permit trading systems, regional renewable energy credit trading systems, penalties of failing to meet emission standards, and projecting the cost of achieving target emissions reductions. The strategies for increasing this benefit are similar to the GHG emission strategies; however, there are already regulations in place for many states.

Implicit Value of PV

This value is driven by customers' willingness to pay a premium price for electricity from a PV system. For some commercial customers, this value could come from demonstrating to their key stakeholders (customers, investors, employees, regulators) that the organization is environmentally friendly. For some residential customers, this value could come from a desire to reduce their environmental impact, create an image of being environmentally friendly, and/or create an image of being an early adopter of emerging technologies. The magnitude of this value across market segments is still unclear. Furthermore, this value may change over time as PV penetrates the market. The implicit value may decline as PV becomes more common. Or conversely, PV may become a "must-have" product for some sectors of the economy. Understanding what creates this value and how it will change over time will be critical to the success of PV.

Equipment and Installation Cost

This value is driven by three key factors: system size, location, and projected long-term costs (i.e., financing). Financing is an important aspect of cost — it varies depending on size and length of payments, interest rates, etc. A typical levelized cost for a residential retrofit system is 29.26 cents/kWh, while the cost for a typical commercial retrofit system is 26.49 cents/kWh. Large systems have a lower cost per output unit than smaller systems as some PV system costs (e.g., design, engineering, transportation, installation, permitting, and incentive request) are mostly fixed. The location is also a factor as labor rates in some regions are more expensive than others, driving up the labor-intensive costs (e.g., design, engineering, and installation).

As the industry continues to grow and mature, economies of scale and learning curves across the supply chain are expected to reduce overall system costs. However, it is still unclear exactly how much costs will come down, and projections have significant variance. A strategy to reduce the cost from this value is to continue to promote incentives and remove regulatory and market barriers that will help the industry grow and achieve the economies of scale. Another strategy is to help capture and disseminate operational best practices from Europe (Germany) and Asia (Japan) across the supply chain that will accelerate.

PV Value Case Studies and Scenarios

We developed a variety of case studies that allowed for a consistent comparison of PV values for specific PV systems. Case studies in Texas, California, Minnesota, Wisconsin, Maryland, New York, Massachusetts, and Washington were reviewed. Seven of these case studies were residential systems and five were commercial systems. The case that had the most information was the Austin Energy study. Below are the results of the study for a 5-kilowatt (kW) residential system.[21] As table E-3 illustrates, customer participants have a positive net present value (NPV) while utilities/ratepayers and society have a negative NPV in this example. It is important to note that in this case study, the central power capacity cost is low, compared to other studies. This is because the displaced marginal resource was a baseload gas turbine plant, while other studies consider peaking plants with limited hours of annual operation as the displaced marginal resource. Another value that is low compared to other studies is the T&D cost savings because load growth is mostly occurring in suburban areas with a relatively low T&D upgrade budget. Values that are taken from the case (either directly or indirectly) are highlighted in color.

Table E-3. PV Values for a 5-kW Residential System in Austin Energy Territory

	PV Values	Customer/ Participant (¢/kWh)	Utility/ Ratepayers (¢/kWh)	Society (¢/kWh)	Net (¢/kWh)
Benefits	Central Power Generation Cost	-	7.0	-	7.0
	Central Power Capacity Cost	-	1.1	-	1.1
	T&D Costs	-	0.1	-	0.1
	System Losses	-	0.6	-	0.6
	Ancillary Services		0.8		0.8
	Hedge Value		0.0*		0.0
	Customer Price Protection		0.8		0.8
	Criteria Pollutant Emissions				
	Greenhouse Gas Emissions	-	-	2.0**	2.0
	Implicit Value				
Costs	Equipment and Installation	(29.3)	-	-	(30.6)
	PV O&M Expenses	(0.1)	-	-	(0.1)
	Benefits Overhead	(0.2)			(0.2)
Transfer	PV Owner Electricity Bill	6.3	(6.3)	-	-
	Federal Incentives	1.6	-	(1.6)	-
	State Incentives	0.0	-	0.0	-
	Stakeholder Total	(20.9)	3.3	0.4	(17.2)

Note: Values with negligible amounts excluded. Future values discounted at 8.25%. Source: NCI analysis; The Value of Distributed Photovoltaics to Austin Energy and the City of Austin, Clean Power Research.

Although there was incomplete information in each of the cases, we were able to use our model to provide reasonably accurate values for the missing information. In addition to

---

[21] Hoff, T.E., Perez, R., Braun, G., Kuhn, M., Norris, B., The Value of Distributed Photovoltaics to Austin Energy and the City of Austin, Clean Power Research LLC (March 17, 2006)

the case studies, we created six scenarios that demonstrate the model capability. The scenarios are a PV system with storage, a PV system with demand response, a low installation and equipment cost scenario, a low installation and equipment cost scenario with no incentives, a $30/ton GHG scenario and a $50/ton GHG scenario.

PV Value R&D Recommendations

It is recommended that NREL and DOE enhance their efforts to fund R&D that will increase the magnitude and clarity of value from grid connected PV systems. More specifically:

Over the short term

- Promote a standard framework and develop tools easily available to industry to assess the value of PV systems.
- Take a leadership role in the development of a standard approach to value GHG and criteria pollutant emissions.
- Quantify the costs and benefits associated with integrating current and emerging energy storage systems and demand response applications with PV systems.

Over the midterm

- Collaborate with utilities in the development and deployment of new technologies and operating practices to increase the value captured by utilities and ratepayers from PV systems.

Over the long term

- Take a leadership role in establishing frameworks for long-term policies, regulations, and incentives that will reduce the risks and uncertainty currently limiting investment in PV markets.

To develop a standard framework and tools to quantify the value of PV systems, DOE should look at existing programs assessing the value of PV, such as the ones sponsored by the Massachusetts Technology Collaborative (MTC), Sacramento Municipal Utility District (SMUD), and Austin Energy, and identify the next R&D steps for those programs. For example, the MTC is undertaking a pilot program to place a large amount of PV within a congested area of the grid to test the actual values of PV. DOE might consider offering to review the results and leverage some dollars to support the analysis of results and take next steps in funding follow-on R&D. DOE can also support additional analyses of PV values within utility distribution networks, which the California Energy Commission is currently funding. In these cases, there were areas of contention, such as GHG emissions, the implicit value of PV, health benefits of emissions, T&D capacity values, and storage and demand response that could benefit from additional analysis. Some studies had a conservative view of benefits while others had high expectations for the future.

In addition, DOE should collaborate with utilities in the development and deployment of new technologies and operating practices to increase the value captured by utilities and ratepayers from PV systems. These research efforts would help improve the value of grid connected PV systems, reach economic parity with traditional central generation options, and increase their market penetration. Table E-4 provides the most critical R&D recommendations.

Table E-4. Selected PV Value R&D Recommendations

PV Value	R&D Opportunities
Central Power Capacity Cost	• Generally improve the cost and performance of energy storage systems • Understand how demand response can help PV systems displace central capacity
Central Power Generation Cost	• Assess actual performance of installed PV systems against predicted performance
T&D Costs	• Develop cost-effective monitoring, control and validation for performance and physical assurance • Develop tools for utilities to incorporate PV system output uncertainty into T&D planning and capacity deferment processes • Research technical and economical feasibility to use smart grids to manage bidirectional power flows from PV systems and plug-in hybrids
Greenhouse Gas Emissions	• Launch a consensus based effort to standardize emissions valuation method.
Implicit Value of PV	• Use surveys/focus groups to understand motivations within each stakeholder group (e.g., consumers, developers, utilities) behind PV projects that went ahead even with a negative net present value
System Resiliency	• Collect data from utilities or on a region-wide basis regarding number of hours of power outage and assess which outage hours could have been avoided by the use of PV with distributed energy storage
Customer Reliability	• Assess the value of backup power for critical loads that can be provided by PV with distributed energy storage for residential and commercial customer classes
Customer Electric Price Protection	• Develop an accepted methodology to value the certainty of fuel prices
Equipment and Installation Cost	• Compare technology requirements in Europe, Japan and the U.S. and assess the impact of codes on PV system design and costs
Benefits Overhead	• Understand what data and infrastructure (e.g., meters, data management, remote controlling and central monitoring) is required, and what it will take for utilities to rely on PV as a resource

1.0. INTRODUCTION

1.1. Background and Context

Due to accelerated cost reductions and associated growth in production, renewable energy technologies such as solar photovoltaics (PV) and wind are expected to grow rapidly in the United States during the next couple of decades. As these technologies mature they have the potential to provide a significant share of our nation's electricity demand.

However, as their market share grows, concern about potential impacts on the stability and operation of the electricity grid may create barriers to their future expansion. To overcome these potential barriers, the U.S. Department of Energy (DOE) launched the *Renewable System Integration (RSI) Study*. This study will address both the technical and analytical challenges that need to be tackled to enable high penetration levels of solar, wind, and other renewable energy technologies. Of particular interest are approaches that include comprehensive analyses that were used for decision making, rather than an academic study, because the ultimate goal is for stakeholder groups to make decisions on solar investment based on their understanding of the value to them.

By combining renewable technologies with storage, controls, and other appropriate technologies, the RSI Study will build the foundation for realizing a high penetration renewable energy future while enhancing the operation of the electricity grid. In addition, by directly engaging utilities and other stakeholders in this process, this study will build the confidence of regulators and utilities with respect to maximizing the use of renewable energy technologies.

Integrating renewable energy into the grid consists of two distinct elements, centralized renewable generation and distributed renewable generation. This study focuses on the distributed generation element that starts with solar photovoltaic (PV) technology that interconnects at the distribution level (less than 15 kV) to facilitate the widespread market penetration of renewable energy technologies, including storage systems, advanced power electronics, and controls into the U.S. electricity grid. The RSI study included 14 reports. This document represents one of the RSI reports, focused on quantifying the value of grid-connected PV.

1.2. Study Objectives and Scope

The purpose of the *Value Analysis* is to integrate the results of previous research on the value of distributed PV and define needs for additional R&D. The chapter covers the following information:

- The various potential values of distributed PV, including the cost and benefits to participating customers, utilities/ratepayers, and society
- The best methodologies used for estimating key PV values
- A base case quantification and range for each key PV value
- The gaps in existing knowledge and corresponding R&D recommendations • Case studies demonstrating value of distributed PV.

In addition to this chapter, a simple Microsoft® Excel-based tool has been developed to conduct sensitivity analyses around each key value.

2.0. PROJECT APPROACH

The project was divided into eight major tasks:

- Task 1: Analyze existing research PV values
- Task 2: Carry out case studies for PV values
- Task 3: Define PV values
- Task 4: Quantify ranges for PV values
- Task 5: Identify key value drivers
- Task 6: Identify gaps to increase value from PV
- Task 7: Define R&D recommendations
- Task 8: Develop a simple Excel tool.

In Task 1, a comprehensive examination of the previous research related to the value of PV was performed. We included research reports from a wide range of sources, including internal NCI databases, NREL's PV Value Clearinghouse database, and other databases available on the internet. While some reports took a comprehensive and holistic approach and quantified a broad set of values, other reports focused on a single value and provided greater research depth for that single value. Section 8.0, References, lists the research reports that were reviewed for this chapter.

In Task 2, we identified existing case studies that quantified the value of a specific PV system installation. In addition to the sources listed above, we also looked for case studies with leading utilities, equipment manufacturers, and project developers. Unfortunately, most of the case studies quantified the PV values for a single stakeholder. In most cases, the stakeholder was either the participant customer installing a PV system on a roof or a utility that had a customer installing a PV system. We selected case studies with the most complete set of information and used our model to fill in missing values.

In Task 3, we integrated the previous research on the value of PV and defined a list of 19 PV values. This list provides a mutually exclusive and collectively exhaustive framework to assess the benefits and costs associated with grid connected PV systems. The preliminary list of values was presented to the RSI team during the July 26, 2007, interim presentation meeting. Input was received and incorporated into the final list of PV values.

In Task 4, we integrated existing research that quantified the PV values. Different methodologies used to quantify each of the values were documented. The range of magnitudes assigned to each value by the different research reports was also documented. The value ranges were assigned to key stakeholder groups (participant customer, utility/ratepayers, and society) either as a positive financial impact (benefit) or a negative financial impact (cost). The PV values with net positive financial impact across key stakeholders were classified as benefits, the PV values with net negative financial impact were classified as costs, and the PV values with net zero financial impact were classified as transfers.

In Task 5, we summarized the key value drivers of PV. This task was performed by analyzing values with the highest magnitude and defining the variables that impacted each of those values. The similarities among the variables of the high magnitude values provided the bases for the key drivers.

In Task 6, we identified gaps to improve the value from PV systems. This task focused on finding opportunities to increase the benefits or reduce the costs of the highest magnitude PV values. Several strategies were identified for each of the highest magnitude PV values.

In Task 7, we defined opportunities for DOE and NREL to sponsor additional R&D efforts to improve the quantification of PV values and/or improve the value captured from PV systems. With a few exceptions, most of the high magnitude PV values had well established methodologies to quantify the financial impacts, while some of the PV values with lower magnitude PV values do not have generally accepted methods. As a result, recommendations for improving the quantification centered on lower magnitude values. However, almost all values provided an opportunity to increase the benefits and reduce costs through additional R&D. A preliminary list of R&D opportunities by PV value was identified.

In Task 8, we developed a simple Excel tool to estimate the value of PV under different scenarios. The tool is based on the resulting quantification of values from existing research reports that was documented in Task 4. The user inputs the utility that provides service to the customer installing the PV system, the type of customer, and a highmedium-low scenario for several of the PV values. The tool then creates an output table that quantifies the benefits and costs associated with each PV value to each of the key stakeholder groups (i.e., participant customer, utility/ratepayers, and society).

A base case study was performed for both a representative residential and commercial Texas PV system. The purpose is to demonstrate the value of employing a comprehensive approach for assessing the value of distributed PV. The study also helps to understand and quantify the value that has led stakeholders to carry out projects with a negative NPV.

3.0. CURRENT RESEARCH ON VALUE OF PV

3.1. Description of PV Values

We began with a comprehensive list of quantitative and qualitative PV benefits and costs. After further discussion and collaboration with key stakeholder groups and industry experts, the list was narrowed for the purpose of this chapter to 19 mutually exclusive and collectively exhaustive values of distributed PV (table 1).

Table 1. Key Values of Distributed PV

PV Value	Description
Central Power Generation Cost	PV systems produce electricity, reducing the amount of electricity that needs to be generated at other plants, which in turn decreases fuel and other O&M costs.
Central Power Capacity Cost	PV indirectly avoids and/or defers central power plant capacity investments by reducing demand☐side consumption. Generation capacity value is the economic value of the avoided and/or deferred incremental resource (typically natural gas turbine) reflecting PV's peak load reduction. NIMBY opposition and higher construction costs are driving capacity costs up. PV also avoids the cost of running more expensive plants during peak loads.
T&D Cost	PV avoids and/or defers transmission and distribution capacity investments by reducing demand☐side consumption. Transmission and distribution capacity value is the economic value of the avoided and/or deferred incremental resource reflecting PV's peak load reduction. NIMBY opposition and higher construction costs are driving capacity costs up. This value is also applicable to situations involving significant congestion issues.
System ^{Losses}	Avoided electric system losses are an indirect benefit because they increase the value of other benefits including energy production, generation capacity, environmental and T&D capacity.
Ancillary Services	Utilities can use inverters in PV systems to provide reactive power back to the grid. This increases power quality and could avoid the installation of capacitors.
System Resiliency	Significant deployment of PV systems coupled with storage could provide disaster recovery benefits.
Hedge Value	Current electricity generation is heavily dependant on natural gas and coal. Recent environmental constraints suggest that utilities will become more dependent on natural gas. PV lessens the exposure of the utility to volatile fuel prices and provides stable and predictable electricity prices.

PV Value	Description
Market Price Impacts/Elasticity	The elasticity of demand for electricity supply increases with more PV. Increased demand for PV may decrease the price of electricity from PV, spur market development, thereby further reducing the cost of PV. A decrease in the cost may then increase the demand for this lower cost good.
Customer Electricity Price Protection	Since there is no fuel expense, the costs of electricity from PV will not increase over the life of the system due to fuel costs and the consumer effectively locks in an electricity price.
Customer Reliability	PV can provide electricity to the PV owner during outages because it is not dependent on the grid. The electricity during an outage is limited to sunlight availability. Storage systems could help offset the intermittency issue and increase the reliability value to the owner.
Criteria [Pollutant] Emissions	PV systems eliminate criteria pollutant emissions (e.g., NO_x, SO_x) associated with non□renewable generation resources. Health benefits associated with reduced emissions are included in this value.
Greenhouse Gas Emissions	PV systems eliminate greenhouse gases (CO2) associated with non□renewable generation resources. Renewable Portfolio Standards (RPS) and Renewable Energy Certificates (REC's) are common mechanisms to value emission reductions from renewable sources of energy such as PV.
Implicit Value of [PV]	The intrinsic societal value of PV to customers (e.g., environmental friendliness, feeling good, early adopter) and utilities (e.g., public relations, regulator compliance).

3.2. Drivers of PV Value

The two main drivers for the highest magnitude values are location of the PV system and timing of the power output of the system. As illustrated in Figure 1, a PV system will have higher benefits when it is located in a highly congested distribution system, where there is high insolation to increase production of the PV system, and where gas prices are high. PV systems will also have higher benefits when a large share of their production is during peak demand periods, and when PV systems displace expensive peaking plants that have lower efficiency and utilization, and use more expensive fuel.

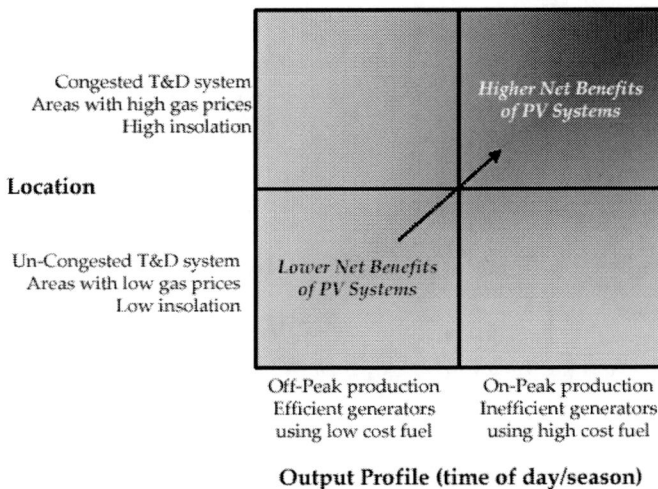

Figure 1. Key Drivers of PV Value.

3.3. Quantification of PV Values

PV values were quantified and allocated to several categories of stakeholders: customer participant, utilities/ratepayers and society. Customer participants take the perspective of PV system owners and end-users. Utilities/ratepayers represent all electric utility customers in the region. Finally, society represents the general population.

We defined a range for each value, because PV values have multiple drivers that can cause values to vary across the United States. The high end of the range corresponds to the high values found in the literature search. Our perception is that this value corresponds to the 90th percentile because there will be isolated situations where true values may be even higher. In a similar fashion, the low end of the range corresponds to the low end of values found in the literature search, representing a value in the 10th percentile because of the isolated situations where values may be even lower. A summary of each value and range along with the driving factors for each value is presented below.

Table 2. PV Value Ranges

	PV Values	Customer/ Participant (¢/kWh)	Utility / Ratepayers (¢/kWh)	Society (¢/kWh)	Net (¢/kWh)	Value Drivers
Benefits	Central Power Generation Cost		3.2 – 9.7		3.2 – 9.7	Gas price, heat rate
	Central Power Capacity Cost		1.1 – 10.8		1.1 – 10.8	Effective load carrying capacity factor, gas turbine capital cost, life adjustment
	T&D Costs		0.1 – 10.0		0.1 – 10.0	Location, growth, climate
	System Losses		0.5 – 4.3		0.5 – 4.3	Location, time period, other benefits
	Ancillary Services		0 – 1.5		0 – 1.5	Ancillary service prices, voltage support
	System Resiliency		Low		Low	Quantification methodology unclear
	Hedge Value		0 – 0.9		0 – 0.9	Gas price forecasts, futures, heat rate
	Market Price Impacts/Elasticity		Low		Low	Quantification methodology unclear
	Customer Price Protection	0.5 – 1.0			0.5 – 1.0	Calculation method
	Customer Reliability	Low			Low	Quantification methodology unclear
	Criteria Pollutant Emissions			0.02 – 2.0	0.02 – 2.0	Market value of emissions
	Greenhouse Gas Emissions			0.02 – 4.2	0.02 – 4.2	Reduction costs, market value, discount rate
	Implicit Value of PV			0 – 2.0	0 – 2.0	Customer willingness to pay premium
Costs	Equipment and Installation	(47) – (19)			(47) – (19)	Size, location
	PV O&M Expenses	(0.15) – (0.05)			(0.15) – (0.05)	Type of system
	Benefits Overhead		(0.2) – (0.1)		(0.2) – (0.1)	Infrastructure and administrative costs
Transfer	PV Owner Electricity Bill	1.1 – 33.0	(33.0) – (1.1)		-	Customer type, rate structure, load profile
	Federal Incentives	1.50 – 7.95		(7.95) – (1.50)	-	Customer type, size, cap
	State Incentives	0 – 17.8		(17.8) – 0	-	State, customer type, size, production, cap
	Stakeholder Total	(43.97) – 40.7	(28.3) – 36.0	(25.7) – 6.6	(41.9) – 27.3	

The remainder of this section discusses the drivers for each PV value in more detail and the methodology and assumptions used to calculate each value.

3.3.1. Central Power Generation Cost

Methodology 1: External Market Pricing

Marginal Cost of Electricity Generation (cents/kWh) = Natural Gas Cost for Power Plants ($/MBtu) × 100 × Heat Rate for Natural Gas Power Plants (Btu/kWh) / 1,000,000 + O&M Costs (cents/kWh).[22]

The methodology for this value uses the avoided cost of natural gas (NG) and in some cases variable O&M costs to determine avoided generation costs. NG costs can be derived by multiplying NG futures contract prices on the New York Mercantile Exchange (NYMEX) by an assumed heat rate. The NYMEX range ($4.570 to $8.752/MBtu) is used extensively for NG prices.[232425]

The Vote Solar White Paper[26], a study consisting of various service regions in California, uses a different approach. At the high end of the range the price of NG during the California energy crisis ($8.05/MBtu) is used, while at the low end the average NYMEX price ($6.49/kWh) is used. In most studies, heat rates in the range of 7,100 Btu/kWh to 11,100 Btu/kWh are used. In the Navigant MTC report, it was found the average heat rate is $7,000 Btu/kWh.[27] The Vote Solar White Paper uses a range of heat rates that vary depending on the age of the power plant fleet and the time of use. Non-peak power plants vary from 8,740 Btu/kWh to 9,690 Btu/kWh while a nominal rate of 9,390 Btu/kWh is used for peak power plants.

The Duke, et al. study[28] estimates O&M costs at less than a cent/kWh while the Vote Solar white paper estimates O&M costs to be 1.1 cents/kWh. The ASPv report values avoided variable O&M costs from 0.01 cent/kWh to 0.14 cent/kWh. The Vote Solar White Paper values avoided natural gas costs from 5.6 cents/kWh (PG&E) to 6.3 cents/kWh (SDG&E & SCE) for non-peak generation under the electric utility scenario (9.5% cost of capital with 20-year recovery period). Non-peak, generation-avoided natural gas values range from 7.6 cents/kWh to 8.5 cents/kWh under the merchant power plant developer scenario (15% cost of capital with 10-year recovery period). The on-peak generation values are 6.6 cents/kWh under the electric utility scenario and 8.0 cents/kWh under the merchant power plant scenario. (These values did not vary depending on the service region.) The range for the Rocky Mountain Institute[4] study, which also analyzes the California region, is 6.5 cents/kWh to 8.0 cents/kWh. The ASPv report values the avoided cost of natural gas from 3.2 cents/kWh to 9.7 cents/kWh.

Some reports simply use a utility's internal marginal cost to forecast its avoided central power plant generating cost. The marginal cost is usually adjusted to reflect current natural

---

[22] Futures price data can be obtained at http://www.nymex.com/markets/newquotes.cfm and other natural gas information can be obtained at http://tonto.eia.doe.gov/oog/info/ngw/ngupdate.asp.

[23] Americans for Solar Power (ASPv), *Build-Up of PV Value in California* (April 13, 2005)

[24] Energy and Environmental Economics, Inc. and Rocky Mountain Institute, *Methodology and Forecast of Long Term Avoided Costs for the Evaluation of California Energy Efficiency Programs* (October 25, 2004)

[25] Itron, Inc. *CPUC Self-Generation Incentive Program Preliminary Cost-Effectiveness Evaluation Report,* (September 14, 2006)

[26] Smellof E., Quantifying the Benefits of Solar Power for California (January 2005)

[27] Navigant Consulting Inc., Distributed Generation and Distribution Planning: An Economic Analysis for the Massachusetts DG Collaborative (February 12, 2006)

[28] Duke, Richard, Robert Williams and Adam Payne, Accelerating Residential PV Expansion: Demand Analysis for Competitive Electricity Markets (2004)

gas prices (using NYMEX). The Austin Energy Report[29] and the SMUD Report[30] use this approach. The Austin Energy Report yielded a range of 7.0 cents/kWh to7.2 cents/kWh.

Table 3. Range and Drivers: Central Power Generation Cost

Range of Value	Net ($\square$/kWh)	Drivers
High End of Range (90% percentile)	9.7	NG Price variance: The high end of the range of NG futures contract prices on the NYMEX ($8.75/MMBtu), the California energy crisis average of $8.05/MMBtu[1]. Prices also vary by region $\square$ MA had a price of $9.10/MMBtu in 2005[2]. The low end of natural gas futures contract prices on the NYMEX ($4.57/MMBtu) is generally used as the low end of the range.
Low End of Range (10% percentile)	3.2	Heat Rate variance: Different areas have different heat rates depending on the age of the fleet –9,720 Btu/kWh vs. 8,740 Btu/kWh in the California area compared to a newer fleet in MA that has an average heat rate of 7,000 Btu/kWh[1,2]. Capital Cost Recovery Factor variance: high end is merchant power plant (15% capital cost over a 10 yr. period) and low end is electric utility (9.5% capital cost over 20 yr. period)1.

1 Smellof E., Quantifying the Benefits of Solar Power for California (January 2005) http://www.votesolar.org/toolsQuantifyingSolar%27sBenefits.pdf
2 Navigant Consulting Inc., Distributed Generation and Distribution Planning: An Economic Analysis for the Massachusetts DG Collaborative (February 12, 2006)
3 Hoff, T.E., Perez, R., Braun, G., Kuhn, M., Norris, B., The Value of Distributed Photovoltaics to Austin Energy and the City ofAustin, Clean Power Research LLC, (March 17, 2006) http://www.austinenergy.com/About%20Us/Newsroom/Reports/PV$\square$ValueReport.pdf

3.3.2. Central Power Capacity Cost

Methodology 1: Economic and Technical Analysis
Central Power Plant Capacity (cents/kWh) = Capital Cost ($/kW) × 100 × ELCC Factor × Levelization Factor

Levelization Factor = Discount Factor x Annual Energy

The levelization factor is used to convert cents/kW to cents/kWh and is critical in the analysis. The discount rate and hours of energy generation must be estimated. The economic analysis involves determining a capacity value, which is assumed to be either a gas peaking turbine or a combined-cycle gas turbine in most studies. The capital costs vary

---

[29] Hoff, T.E., Perez, R., Braun, G., Kuhn, M., Norris, B., The Value of Distributed Photovoltaics to Austin Energy and the City of Austin, Clean Power Research LLC, (March 17, 2006)

[30] Hoff, T.E, Final Results Report with a Determination of Stacked Benefits of Both Utility-Owned and Customer-Owned PV Systems, Clean Power Research, L.L.C (2002)

from \$419/kW to \$619/kW[31]. Values for other reports fell into this range. The effective capacity of PV is the effective contribution it provides to the available generation capacity of the utility. The Effective Load Carrying Capacity (ELCC) factor is widely used to determine how much of the PV capacity can contribute to alleviate utility peak loads. This factor is dependant on the region and orientation of the photovoltaic cells. The Vote Solar White Paper[32] uses an ELCC factor of 50% for the region of California while the ASPv Report[33] uses 65%. The Austin Energy Report uses an ELCC range of 47% to 62% depending on the orientation of the PV cells[34]. A study conducted by NREL shows 20 utility service areas have matches to load shapes ranging from 3 6-70%. The range for the Vote Solar White Paper[13] was 6.2 cents/kWh to 10.8 cents/kWh.

Methodology 2: Life Adjustment Factor

Central Power Plant Capacity (cents/kWh) = Capital Cost (\$/kW) × 100 × ELCC Factor × Levelization Factor × Life Adjustment Factor

Life Adjustment Factor = [GT Life (yrs) − PV Life (yrs)]/[GT Life (yrs)] × (1/(1+Discount Rate)^25)

This method includes a life adjustment factor. It accounts for the fact that on average the life of a gas turbine is 40 years and the life of a PV system is 30 years. In the Austin Energy Report a 7% discount factor is used, and a 25% salvage value is used for the GT, resulting in a 3.3% longer life for the GT[14]. The range for the Austin Energy Report[35] is 1.1 cents/kWh to 1.8 cents/kWh.

Table 4. Range and Drivers: Central Power Capacity Cost

Range of Value	Net (¢/kWh)	Drivers
High End of Range (90% percentile)	10.8	ELCC Factor: High end occurs when peak load coincides with high solar availability. NREL conducted an analysis for 20 utilities and found 70% at the high end[1] – ASPv used 65%[2] while Ed Smellof was close to the low end at 50%. Orientation of PV: The AE Report looked at how different orientations affected the ELCC and found a range of 47% (west at 45 deg) to 62% (1 axis at 30 deg)[3].
Low End of Range (10% percentile)	1.1	Capital Cost of NG Turbine: Varies depending on region, i$_n$ California it was assumed to be \$5 15/kW while in Austin it was assumed to be \$475. Life Adjustment Factor: Most studies did not account for this except for the Austin Energy Report. This increased the value of the gas turbine by 1.033.

1 Smellof E., Quantifying the Benefits of Solar Power for California (January 2005) http://www.votesolar.org/tools QuantifyingSolar%27sBenefits.pdf

---

[31] Americans for Solar Power (ASPv), *Build-Up of PV Value in California* (April 13, 2005)

[32] Smellof E., Quantifying the Benefits of Solar Power for California (January 2005)

[33] Americans for Solar Power (ASPv), *Build-Up of PV Value in California* (April 13, 2005)

[34] Hoff, T.E., Perez, R., Braun, G., Kuhn, M., Norris, B., *The Value of Distributed Photovoltaics to Austin Energy and the City of Austin*, Clean Power Research LLC, (March 17, 2006)

[35] *Ibid*

2 Americans for Solar Power (ASPv), Build-Up ofPV Value in California (April 13, 2005)
   http://www.mtpc.org/renewableenergy/public policy/DG/resources/2005-04-CA-PV-Value-Links-
   R04-03-017.pdf
3 Hoff, T.E., Perez, R., Braun, G., Kuhn, M., Norris, B., The Value of Distributed Photovoltaics to
   Austin Energy and the City ofAustin, Clean Power Research LLC, (March 17, 2006)
   http://www.austinenergy.com/About%20Us/Newsroom/Reports/PV-ValueReport.pdf

3.3.3. T&D Costs

Methodology 1: T&D Growth Method
    Deferred T&D Capital Cost Value (cents/kWh) = [(Cost of T&D Investment Plan ($)
× 100 × Value of Money (%) × ELCC Factor)/Load Growth] × Levelization Factor

Although this value includes both transmission and distribution, there are cases that are
specific to one or the other. For example, PV can be installed in an area that reduces the
transmission peak but is in a distribution area with excess capacity, thus providing
limited value. The cost of the T&D investment plan comprises a number of different values
such as feeder upgrade deferral, transformer lifetime increase, reduced load-tap
changer/voltage regular use, etc.

Deferred T&D costs depend heavily on the ELCC factor and growth in the region.
The ELCC method is used extensively in studies. In the Austin Energy Report[36] the ELCC
factor varied between 46% and 63% depending on the orientation, while the Rocky Mountain
Institute Report[37] and the Vote Solar White Paper[38] used nominal rates of 65% and 50%
respectively. Annual growth ranges from 0.5 megawatts per year (MW/yr) to 12.3 MW/yr
for the Austin Energy Report, while the Vote Solar White Paper assumes a constant growth rate
independent of the region of 5%.

Ideal resource values range from $3 9/kW to $445/kW for the Austin Energy Report
based on region. The Rocky Mountain Institute Report also breaks up the costs by climate
zone and region, from $39/kW to $88/kW for the Vote Solar White Paper and from
$9/kW to $88/kW for the Rocky Mountain Institute Report. The Austin Energy Report
yields a levelized 'cost range of 0.1 cents/kWh to 0.2 cents/kWh. The Rocky Mountain
Institute White Paper report yields a levelized cost range of 4.5 cents/kWh to 10
cents/kWh. The high value generally applies to situations where there are high T&D costs with
good load match.

---

[36] Hoff, T.E., Perez, R., Braun, G., Kuhn, M., Norris, B., The Value of Distributed Photovoltaics to Austin Energy and
     the City of Austin, Clean Power Research LLC, (March 17, 2006)
[37] Energy and Environmental Economics, Inc. and Rocky Mountain Institute, Methodology and Forecast of
     Long Term Avoided Costs for the Evaluation of California Energy Efficiency Programs (October 25, 2004)
[38] Smellof E., Quantifying the Benefits of Solar Power for California (January 2005)

Table 5. Range and Drivers: T&D Costs

Range of Value	Net ($\square$/kWh)	Drivers
High End of Range (90% percentile)	10	Location: Urban areas, such as downtown Austin ($445/kW)[1], have more expensive T&D upgrade costs, whereas areas such as rural California have cheaper T&D upgrade costs ($9/kW)[2]. Temperature: Regions where temperature spikes dramatically, T&D avoided costs are high, whereas in coastal regions with moderate temperatures, T&D costs are lower (i.e. East Bay - $9 .25/kW)[2]
Low End of Range (10% percentile)	0.1	Growth: Projected load growth varies. In downtown Austin growth is small (0.4 MW/yr) while in the Southwest it's 12.3 MW/yr.[1] Location & Growth: Areas where there is high growth and high costs of T&D upgrades are at the high end, whereas areas where they don't coincide (such as Austin) are on the low end.

1 Hoff, T.E., Perez, R., Braun, G., Kuhn, M., Norris, B., *The Value of Distributed Photovoltaics to Austin Energy and the City of Austin*, Clean Power Research LLC, (March 17, 2006) http://www.austinenergy.com/About%20Us/Newsroom/Reports/PV$\square$ValueReport.pdf
2 Energy and Environmental Economics, Inc. (E3) and Rocky Mountain Institute, *Methodology and Forecast of Long Term Avoided Costs for the Evaluation of California Energy Efficiency Programs* (October 25, 2004) http://www.ethree.com/cpuc/E3AvoidedCostsFinal.pdf
3 SmellofE., *Quantifying the Benefits of Solar Power for California* (January 2005) *http://www.votesolar.org/tools QuantifyingSolar%27sBenefits.pdf*

3.3.4. System Losses

Methodology 1: Loss Factors
Avoided Costs with Line Losses (cents/kWh) = (Avoided Generation Capacity Costs + Avoided Energy Production Costs + Avoided T&D Costs + Avoided Environmental Costs) × (Loss Factor -1)

PV reduces system losses by producing power at the point of consumption – it is an indirect benefit because it magnifies the value of other benefits. Methodology 1 uses an approach by performing the calculations twice – with and without loss impacts. Taking the difference between the two provides an explicit value for this benefit.

Studies such as the Vote Solar White Paper[39] estimate loss factors for each benefit. The Austin Energy Report[40] describes three different methods of calculating the loss factor: calculating the system average (Total Electricity Loss/Total Electricity Produced), the incremental change in losses that occur during peak loads and the incremental changes in losses that occur over all hours in the time period. The latter is the method used by the

---

[39] Smellof, E., Quantifying the Benefits of Solar Power for California (January 2005)

[40] Hoff, T.E., Perez, R., Braun, G., Kuhn, M., Norris, B., The Value of Distributed Photovoltaics to Austin Energy and the City of Austin, Clean Power Research LLC, (March 17, 2006)

SMUD Report[41]. The Navigant MTC Report found that electrical system losses in the T&D system alone are 2% to 6%[42]. In some studies all of the categories are grouped together under one factor. In the Austin Energy Report the various categories were split up. The report found that the greatest loss savings were in energy production followed by generation capacity, environmental, and T&D. In the Austin Energy Report the loss savings benefit ranged from 0.6 cents/kWh to 0.7 cents/kWh. The Vote Solar White Paper has a range of 0.7 cents/kWh to 2.6 cents/kWh while the ASPv report has a range of 0.52 cents/kWh to 1.36 cents/kWh. The Duke et al: Report has a loss value of 4.3 cents/kWh.

Methodology 2: Credit
This is an implicit method that credits the PV system to account for the reduction in losses. Each value calculation is performed using the higher kW or kWh figures. None of the studies investigated use this method because it does not provide an explicit number.

Table 6. Range and Drivers: System Losses

Range of Value	Net ($\square$/kWh)	Drivers
High End of Range (90% percentile)	4.3	System Location: The loss factor is dependant on the location. The further away from the power plant, the higher the loses. Some utility systems have higher average loses than others (i.e. SDG&E 8% vs. PG&E 9%)[1].
Low End of Range (10% percentile)	0.5	Time Period: The magnitude of loss factors is dependant on the time period that electricity is produced (i.e. peak vs. non-peak). SDG had 11% during peak and 8% during non-peak.[1] Type of Benefit: The impact of loss savings varies for each value. Generation savings are impacted the most (i.e. 9% vs. T&D 7.4%) 2

1 SmellofE., Quantifying the Benefits of Solar Powerfor California (January 2005) http://www.votesolar.org/toolsQuantifyingSolar%27sBenefits.pdf
2 Energy and Environmental Economics, Inc. (E3) and Rocky Mountain Institute, *Methodology and Forecast of Long Term Avoided Costs for the Evaluation of California Energy Efficiency Programs* (October 25,2004) http://www.ethree.com/cpuc/E3AvoidedCostsFinal.pdf

3.3.5. Ancillary Services

Methodology 1: Utility Bill Analysis
Ancillary Services include: VAR Support, load following, operating reserves, and dispatch and scheduling. The distributed generation (DG) units are unlikely or unable to participate in the markets for load following, operating reserves, and dispatch and scheduling. Although unlikely to participate in the market, synchronous DG may provide some of these services when operating. The potential value of ancillary services to other electric

---

[41] Hoff, T.E, Final Results Report with a Determination of Stacked Benefits of Both Utility-Owned and Customer-Owned PV Systems, Clean Power Research, L.L.C (2002)
[42] Navigant Consulting Inc., Distributed Generation and Distribution Planning: An Economic Analysis for the Massachusetts DG Collaborative (February 12, 2006)

ratepayers for PV used in the Rocky Mountain Institute Report[43] is valued at the CAISO market price range of 0.5 to 1.5 cents/kWh. The Vote Solar White Paper[44] values ancillary services at 0.2 cents/kWh. The Austin Energy Report[45] evaluates the voltage regulation benefit by assuming that PV inverters could be modified to operate the desired power factor. The results suggest that although there is a range depending on how much the PV system can be depended on for voltage support, the value will always be close to 0 cents/kWh. The MTC report by NCI values ancillary services at 0.3 cents/kWh, based on the E3 Report.[46]

Table 7. Range and Drivers: Ancillary Services

Range of Value	Net (¢ /kWh)	Drivers
High End of Range (90% percentile)	1.5	Ancillary Service Prices
Low End of Range (10% percentile)	—	Perceived reliability for voltage support.

3.3.6. Hedge Value

Methodology 1: Guarantee Electricity Supply Costs

Natural gas hedge value ($/kWh) = Cost to guarantee that a portion of electricity supply costs are fixed ($/kWh)

The value equals the cost of natural gas futures discounted at the risk-free discount rate. This analysis requires the natural gas price over the life of the PV system and the risk free discount rate associated with each year of the analysis. The Austin Energy Report uses NYMEX natural gas futures prices and the U.S. Treasury Yield Curve for risk free discounts rates. (The London Interbank Offer Rate (LIBOR) could also be used.) The Austin Energy Report had a discount factor of 0.96 in 2007 and 0.27 in 2035. The ASPv report values of the price hedge from 0.4 to 0.9 cents/kWh.

Methodology 2: Selling Risk Free Benefits

Natural gas hedge value ($/kWh) = Price that entity will pay for risk-reduction benefits ($/kWh). Another method mentioned in the Austin Energy Report to value the PV is to sell natural gas futures (or other contracts) in proportion to natural gas savings from PV.

---

[43] Energy and Environmental Economics, Inc. and Rocky Mountain Institute, Methodology and Forecast of Long Term Avoided Costs for the Evaluation of California Energy Efficiency Programs (October 25, 2004)

[44] Smellof E., Quantifying the Benefits of Solar Power for California (January 2005)

[45] Hoff, T.E., Perez, R., Braun, G., Kuhn, M., Norris, B., The Value of Distributed Photovoltaics to Austin Energy and the City of Austin, Clean Power Research LLC, (March 17, 2006)

[46] Navigant Consulting Inc., Distributed Generation and Distribution Planning: An Economic Analysis for the Massachusetts DG Collaborative (February 12, 2006)

Table 8. Range and Drivers: Hedge Value

Range of Value	Net (¢ /kWh)	Drivers
High End of Range (90% percentile)	0.9	Market Stability: A volatile market creates more value for a hedge while stable market prices decrease this value. Heat Rates: Low efficiency heat rates increase the value.
Low End of Range (10% percentile)	0.0	

1. Wiser, R., Mills, A., Barbose, G., Golove, W., *The Impact of Retail Rate Structures on the Economics of Commercial Photovoltaic Systems in California* (July 2007), Lawrence Berkeley National Laboratory http://eetd.lbl.gov/ea/ems/reports/63019.pdf

3.3.7. Implicit Value of PV

Methodology 1: Market Data

Most of the studies analyzed do not have methodologies for determining the intrinsic value of PV. However, the Austin Energy Report[47] examined market data that indicates customer willingness to pay premium prices for green power and found that it had a value of 2 cents/kWh. There are other reports available that analyze green market pricing and EPRI is currently conducting work in this area.

Table 9. Range and Drivers: Implicit Value of PV

Range of Value	Net (¢ /kWh)	Drivers
High End of Range (90% percentile)	2	Customer willingness to pay premium for green power. 1
Low End of Range (10% percentile)	0	

1. Hoff, T.E., Perez, R., Braun, G., Kuhn, M., Norris, B., *The Value of Distributed Photovoltaics to Austin Energy and the City of Austin*, Clean Power Research LLC, (March 17, 2006) http://www.austinenergy.com/About%20Us/Newsroom/Reports/PV-ValueReport.pdf

3.3.8. Market Price Impact/Elasticity

Methodology 1: Long-Term Supply Curve

The Rocky Mountain Institute Report[48] and the California Public Utilities Commission Report[49] were some of the only studies that recognized that reduced demand results in a decrease in the market-clearing price for electricity, and therefore an increase in consumer surplus. However, it was found that this benefit is likely to be small and does not cause a price decrease. Most studies did not try to quantify this value because there are so many factors influencing natural gas demand.

---

[47] Hoff, T.E., Perez, R., Braun, G., Kuhn, M., Norris, B., The Value of Distributed Photovoltaics to Austin Energy and the City of Austin, Clean Power Research LLC, (March 17, 2006)

[48] Energy and Environmental Economics, Inc. and Rocky Mountain Institute, Methodology and Forecast of Long Term Avoided Costs for the Evaluation of California Energy Efficiency Programs (October 25, 2004)

[49] Energy and Environmental Economics, Inc. and Rocky Mountain Institute, Methodology and Forecast of Long Term Avoided Costs for the Evaluation of California Energy Efficiency Programs (October 25, 2004)

3.3.9. Customer Price Protection

Methodology 1: Market Based
Electricity Price Risk Mitigation = ERF– ER (ERF = Cost of Electricity from Risk-Less Generation Asset, ER = Cost of Electricity from Risky Portfolio)

This method consists of calculating the cost of electricity from the risky portfolio and the cost of electricity from the risk-less generation asset and taking the difference. The SMUD Report[50] found this method to yield a premium of 0.5 cents/kWh. (Natural gas- fired generation firmed up with natural gas futures is assumed to be the least cost risk- free generation asset).

Methodology 2: CAPM
Electricity Price Risk Mitigation = (EP– yER)/(1 – y) – ER
(EP = Cost of Electricity from District Portfolio, y = percent of the risky portfolio)

This alternative approach solves for the expected cost of the risk-less asset as shown above. The SMUD Report found this method to yield a premium of 1.0 cents/kWh.

Table 10. Range and Drivers: Customer Price Protection

Range of Value	Net (¢ /kWh)	Drivers
High End of Range (90% percentile)	1	Calculation Method: Method of determining the value of the risk-less generation.
Low End of Range (10% percentile)	0.5	

3.3.10. Customer Reliability
Most of the studies evaluated do not quantify the value of customer reliability. However, several reports by LBNL and NREL look at the benefit of increased outage support for customers using batteries.[51][52] The PV without storage provides little reliability, but with storage it would be worth in the range of 0 to 2.7 cents/kWh depending on the reliability needs of the customer.

3.3.11. Criteria Pollutant Emissions

Methodology 1: Direct Cost Savings
Emissions Benefit ($/kWh) = Market Value of Penalties or Costs ($/kWh)

There are two main methods of assigning an economic value: direct cost savings (penalties or costs to meet standards) and human health benefits. The majority of studies that

---

[50] Hoff, T.E, Final Results Report with a Determination of Stacked Benefits of Both Utility-Owned and Customer-Owned PV Systems, Clean Power Research, L.L.C (2002)

[51] Hoff, T.E., Perez, R., Braun, Margolis, R.M., Maximizing the Value of Customer-Sited PV Systems Using Storage and Controls (2005)

[52] Hoff, T.E., Perez, R., Margolis, R.M., Increasing the Value of Customer-Owned PV Systems Using Batteries (November 9, 2004)

do evaluate the economic benefits of emissions use the direct costs savings method[53][54][55]. The connection between pollution and emission factors (lb/MBtu) is calculated by the U.S. Environmental Protection Agency (EPA) or CEC in most studies. The E3 Report[29] uses South Coast Air Quality Management District (SCAQMD) Reclaim Data, CARB or Utility Planning Documents to come up with Emission Reduction Credits (ERCs) or RECLAIM Trading Credits (RTCs) resulting in an environmental value of 0.7 cents/kWh. The Austin Energy Report[56] does not have an RPS mandate so a typical market-based green power program of 2.0 cents/kWh is used. RPS requirements force electric service providers to either acquire/build renewable-generated electricity or to purchase RECs from others. The SMUD Report[57] uses emissions costs based on the Sacramento Metropolitan Air Quality Management District (SMAQMD) credit banking system which sells emissions permits.

Methodology 2: Health Benefits
Emissions Benefit ($/kWh) = Health Benefits to Society ($/kWh)

This method assigns the avoided health costs and shortened life times due to emissions as the value. This is difficult as the connection between emissions and health is not well established and there are no widely accepted methodologies.[58] The ASPv paper states a range of health benefits from 0.02 cents/kWh to 0.04 cents/kWh, while the Vote Solar White Paper reports a value of 1.4 cents/kWh for NOx emissions. A comprehensive study for the EPA also estimates the health effects (mortality, hospital admissions, bronchitis, work lost days, etc.) due to exposure to particulates from power plants. The report includes detailed methodologies and assumptions, but does not provide an overall levelized cost[59].

Table 11. Range and Drivers: Criteria Pollutant Emissions

Range of Value	Net (¢ /kWh)	Drivers
High End of Range (90% percentile)	0.02	• Market Value (government mandate, geographic region): Some areas place a high market value on emission reductions such as California while and others do not, such as Austin.
Low End of Range (10% percentile)	2.0	

[53] Smellof E., Quantifying the Benefits of Solar Power for California (January, 2005)

[54] Energy and Environmental Economics, Inc. and Rocky Mountain Institute, Methodology and Forecast of Long Term Avoided Costs for the Evaluation of California Energy Efficiency Programs (October 25, 2004)

[55] Hoff, T.E., Perez, R., Braun, G., Kuhn, M., Norris, B., The Value of Distributed Photovoltaics to Austin Energy and the City of Austin, Clean Power Research LLC (March 17, 2006)

[56] Hoff, T.E., Perez, R., Braun, G., Kuhn, M., Norris, B., The Value of Distributed Photovoltaics to Austin Energy and the City of Austin, Clean Power Research LLC, (March 17, 2006)

[57] Hoff, T.E, Final Results Report with a Determination of Stacked Benefits of Both Utility-Owned and Customer-Owned PV Systems, Clean Power Research, L.L.C (2002)

[58] Hoff, T.E., Perez, R., Braun, G., Kuhn, M., Norris, B., The Value of Distributed Photovoltaics to Austin Energy and the City of Austin, Clean Power Research LLC, (March 17, 2006)

[59] ABT Associates and ICF Consulting The Particulate-Related Health Benefits of Reducing Power Plant Emissions, Prepared for Cleat Air Task Force, (October 2000)

3.3.12. Greenhouse Gas Emissions

Methodology 1: Cost-Based Approach

CO2 Emission Benefit ($/kWh) = Emission Intensity (tonnes of CO2/kWh) x (Value of CO2 Emissions ($/tonne)

Although there are several methods to assign value to the GHG emissions (carbon tax, cap-and-trade programs, or REC trading), most studies used compliance costs. The emissions factor varies depending on the type of electricity generation. The E3 Report[60] uses two values to justify a projection of the value of CO2 reductions. One is the short- term price of $5/ton of CO2 based on World Bank, Dutch, and UK market activity. The other is U.S. compliance with Kyoto protocol, which would result in a cost of $17.5/ton of CO2 in 2013.

By discounting the projected stream of the shorter-term $5/ton of CO2 at 8.15% to $17.5/ton of CO2 in 2013, the resulting present value is $8/ton of CO2 (or $29/ton of carbon). This results in a range of 0.33 cents/kWh - 0.52 cents/kWh based on the heat rate range of 7,100 to 11,100 Btu/kWh. The Navigant MTC report found a value of $6.5/ton for "very high emission" projections.[61] A European Commission study detailed the costs of achieving reductions in GHG emissions and found the value to be $66 to $170 per ton of carbon. ASPv and Duke adopted the mid-range of this cost ($1 00/ton of carbon) resulting in a range of 1.1 cents/kWh to 1.8 cents/kWh.[62][63] The Vote Solar Initiative White Paper[64] uses the midpoint ($92/ton of carbon), which results in a benefit of 1.9 cents/kWh. The SMUD Report[65] uses a range from $2.19 (The Climate Trust) to $10 per ton of CO2 (British Petroleum). If the United States passes the Low Carbon

Economy Act, a price of $12 per ton of CO2 will be implemented in 2012. This will result in a range of levelized costs from 0.2 to 1.0 cents/kWh (using Navigant's model of emission intensity rates). The range of values is based on a $1/ton scenario and a $50/ton scenario and the emission intensity per state.

Table 12. Range and Values: Greenhouse Gas Emissions

Range of Value	Net (¢ /kWh)	Drivers
High End of Range (90% percentile)	4.18	Carbon Value: High carbon value yields higher value. Emissions Rates (location/efficiency of plant): Higher carbon emission intensity yield higher value.

[60] Energy and Environmental Economics, Inc. and Rocky Mountain Institute, Methodology and Forecast of Long Term Avoided Costs for the Evaluation of California Energy Efficiency Programs (October 25, 2004)

[61] Navigant Consulting Inc., Distributed Generation and Distribution Planning: An Economic Analysis for the Massachusetts DG Collaborative (February 12, 2006)

[62] Americans for Solar Power (ASPv), Build-Up of PV Value in California (April 13, 2005)

[63] Duke, Richard, Robert Williams and Adam Payne, Accelerating Residential PV Expansion: Demand Analysis for Competitive Electricity Markets (2004)

[64] Smellof E., Quantifying the Benefits of Solar Power for California (January 2005)

[65] Hoff, T.E, Final Results Report with a Determination of Stacked Benefits of Both Utility-Owned and Customer-Owned PV Systems, Clean Power Research, L.L.C (2002)

Table 12. (Continued).

Range of Value	Net (¢ /kWh)	Drivers
Low End of Range (10% percentile)	0.02	

1. Smellof E., Quantifying the Benefits of Solar Power for California (January 2005) http://www.votesolar.org/toolsQuantifyingSolar%27sBenefits.pdf

3.3.13. Equipment and Installation Cost

Methodology 1: Levelized Costs

Cost of Equipment (cents/kWh) = (Capital Cost (cents/kW) + Fixed O&M (cents/kW-yr)) × Levelized Cost Factor

Levelized Cost Factor = Cost of Equipment / (Life of System x DC-AC Conversion Factor x Capacity Factor x Annual Hours)

An NREL assessment uses costs provided by Black and Veatch: $3,600 to $8,050/kW capital costs, $5.7 to $8.2/kW-yr fixed costs resulting in a levelized cost range of 19.4 to 47 cents/kWh. A study by Center for Sustainable Energy (CSE)[66] found that without incentives the cost for a project in New York City was 50.17 cents/kWh. A breakdown of the costs estimates that 60% of the cost is the PV module, 25% balance of system, and 15% system design and installation. It is assumed that costs decrease at a rate of 5%/yr.[67] We estimate current capital costs at 26.49 cents/kWh to 29.26 cents/kWh and project them to decrease to 17.79 cents/kWh to 18.98 cents/kWh by 2015 (in 2006 dollars). Our assumptions were 8,760 operating hours, a capacity factor of 15%, a DC-AC conversion factor of 0.77 and a life of 25 years. We assumed that costs decreased annually by 4% to 6%.

Table 13. Range and Drivers: Equipment and Installation

Range of Value	Net (¢ /kWh)	Drivers
High End of Range (90% percentile)	(19)	Size: Large systems have inherently less cost on a per kW basis than small systems. Location: Labor costs vary significantly in different regions (i.e. higher in NY)[1] Long Term Costs: The amount that costs
Low End of Range (10% percentile)	(47)	

1.Centre for Sustainable Energy at Bronx Community College, *New York City's Solar Energy Future* (January 2006) http://www.bcc.cuny.edu/Institutional Development/CSE/Documents/ CUNY%20MSR%20□%20Market%20for%20PV%20in%20NYC.pdf

---

[66] Center for Sustainable Energy at Bronx Community College, New York City's Solar Energy Future (January 2006)

[67] Del Chiaro, B., Dutzik, T., Vasavada, J., The Economics of Solar Homes in California, Environment California Research and Policy Center (December, 2004)

3.3.14. PV O&M Expenses

Methodology 1: Projected Costs

Most studies assume these costs are negligible. Some other reports assume a variable cost, but it is extremely small (i.e., NREL: 0.05 to 0.15 cents/kWh). Inverters could have an impact and are currently replaced every 7 to 15 years.

Table 14. Range and Drivers: PV O&M Expenses

Range of Value	Net (¢ /kWh)	Drivers
High End of Range (90% percentile)	(0.15)	Type of System
Low End of Range (10% percentile)	(0.05)	

3.3.15. Benefits Overhead

Methodology 1: Comparable Incentive Programs

Benefits Overhead ($) = Benefits Overhead Costs ($/kW) × PV kW

Benefits Overhead Costs are those associated with capturing and monetizing all of the various value streams. This includes program administration and other equipment costs, such as advanced metering and/or the cost to address technical issues for including PV in a distribution deferral solution. The program administration (salaries, facilities, program design, and implementation) and program evaluation costs (for hiring meter installation subcontractors) in California's Self-Generation Incentive Program are $12.5 million. If allocated equally on a per kW basis across all active and complete projects, the cost is equal to $47.75/kW.[68] Using our assumptions, this is converted to 0.2 cents/kWh.

Table 15. Range and Drivers: Benefits Overhead

Range of Value	Net (¢ /kWh)	Drivers
High End of Range (90% percentile)	(0.2)	Rate Structure: Alters the PV value by 25% to 75% depending on size of the PV system relative to the building load[1].
Low End of Range (10% percentile)	(0.1)	Customer Load Profile: Customers that peak in the afternoon can receive significant demand charge savings while facilities with flat or inverted profiles will earn minimal demand charge reduction[1]. PV Production Profile: At 2% penetration, range is $0.10- 0.18/kWh0) while at 75% penetration, range is $0.06- 0.18/kWh) 1

1 Wiser, R., Mills, A., Barbose, G., Golove, W., The Impact of Retail Rate Structures on the Economics of Commercial Photovoltaic Systems in California (July 2007), Lawrence Berkeley National Laboratory http://eetd.lbl.gov/ea/ems/reports/63019.pdf

---

[68] Energy and Environmental Economics, Inc. and Rocky Mountain Institute, Methodology and Forecast of Long Term Avoided Costs for the Evaluation of California Energy Efficiency Programs (October 25, 2004)

3.3.16. System Resiliency

Few studies evaluated reliability in relation to disaster recovery. Currently, with or without storage, a PV system will trip off-line if the grid goes down. However, as engineering and procedural issues improve, PV combined with storage could provide disaster recovery relief. The Austin Energy Report[69] valued this benefit by estimating the disaster cost, disaster propensity, and solar emergency power. Without storage the value was found to be negligible ($0/kWh). The Austin Energy Report suggests that with significant deployment of storage this benefit could have substantial value.

3.3.17. PV Owner Electricity Bill

Methodology 1: Utility Bill Analysis

Electricity Bill Savings ($/kWh) = (Total Bill Without PV − Total Bill With PV)/(Annual Energy PV Production)

Rate structure, system size, orientation and degradation are all factors that impact savings. The Lawrence Berkeley National Laboratory (LBNL) Report[70] shows that the value of PV systems varied immensely across customers and retail rates: 5 to 24 cents/kWh. Our model has a range of 1.1 cents/kWh to 33.0 cents/kWh. The high end value is in California for residential customers on tiered rates. The range in the following table is based on our model.

Table 16. Range and Drivers: PV Owner Electricity Bill

Range of Value	Net (¢/kWh)	Drivers
High End of Range (90% percentile)	33.0	**Rate Structure:** Alters the PV value by 25% to 75% depending on size of the PV system relative to the building load.[1]
Low End of Range (10% percentile)	1.1	**Customer Load Profile:** Customers that peak in the afternoon can receive significant demand charge savings while facilities with flat or inverted profiles will earn minimal demand charge reduction.[1] **PV Production Profile:** At 2% penetration, range is $0.10-0.18/kWh0 while at 75% penetration, range is $0.06-0.18/kWh)[1]

1. Wiser, R., Mills, A., Barbose, G., Golove, W., *The Impact of Retail Rate Structures on the Economics of Commercial Photovoltaic Systems in California* (July 2007), Lawrence Berkeley National Laboratory http://eetd.lbl.gov/ea/emp/reports/63019.pdf

3.3.18. Federal Incentives

Methodology 1: Investment Credits

Federal Incentive (cents/kWh) = Rebate for Cost of Equipment (cents/kWp) x / (Life of System x DC-AC Conversion Factor x Capacity Factor x Annual Hours)

---

[69] Hoff, T.E., Perez, R., Braun, G., Kuhn, M., Norris, B., The Value of Distributed Photovoltaics to Austin Energy and the City of Austin, Clean Power Research LLC, (March 17, 2006)

[70] Wiser, R., Mills, A., Barbose, G., Golove, W., *The Impact of Retail Rate Structures on the Economics of Commercial Photovoltaic Systems in California*, Lawrence Berkeley National Laboratory (July 2007)

Federal law provides a 30% income tax credit of up to $2,000 for purchase and installation of residential PV and a 30% tax credit for commercial systems.[71] At the federal level, there is also 5-year accelerated depreciation. At the residential level, federal incentives generally have a smaller impact than state incentives because of the cap. Thus, residential systems are at the low end, and commercial systems, where there is no cap, are at the high end. We used the following assumptions in our model: 25-year life, DC-AC conversion factor of 0.77, a capacity factor of 15%, and 8,760 annual hours. The range in table 17 is based on our model for federal incentives.

Table 17. Range and Drivers: Federal Incentives

Drivers	Net (¢/kWh)	
High End of Range (90% percentile)	7.95	'Customer segment 'Size of system, as income tax credit is caped
Low End of Range (10% percentile)	1.58	

3.3.19. State Incentives

Methodology 1: Initial Investment or Production Credits
State Incentive (cents/kWh) = Rebate for Cost of Equipment / (Life of System x DC-AC Conversion Factor x Capacity Factor x Annual Hours) + Production (kWh) x Production Credit (cents/kWh)

Incentives vary widely by state. In some states there are no incentives (West Virginia, Wyoming, South Carolina, etc.) while in states such as California they are significant.

In California, the California Solar Initiative (C SI) is managed by the Public Utilities Commission and applies to residential retrofits and commercial and government buildings, and uses a mix of capacity and output based payments. Systems smaller than 100 kilowatt (peak) (kWp) have upfront, expected performance based payments through rate reductions over 10 years. Systems larger than 100 kWp have over time, output based payments through rate reductions over 5 years. The New Solar Home Partnership (NSHP), managed by the California Energy Commission (CEC), applies to new residential housing, residential portions of mixed development, and affordable housing. The incentive program offers buy-downs, in dollars per watt (peak) ($/Wp), corrected for expected performance through a complex calculation that requires an expert site inspection.[72] The range in table 18 is based on our model for state incentives.

---

[71] Itron, Inc. *CPUC Self-Generation Incentive Program Preliminary Cost-Effectiveness Evaluation Report,* (September 14, 2006)

[72] Database of State Incentives for Renewables & Efficiency, www.dsireusa.org

Table 18. Range and Drivers: State Incentives

Range of Value	Net (¢ /kWh)	Drivers
High End of Range (90% percentile)	7.95	• Customer segment • Size of system, as income tax credit is caped
Low End of Range (10% percentile)	1.58	

4.0. PV VALUE CASE STUDIES AND SCENARIOS

4.1. Case Studies

We developed a variety of case studies that allowed for a consistent comparison of PV values for specific situations. Case studies in Texas, California, Minnesota, Wisconsin, Maryland, New York, Massachusetts, and Washington were reviewed. Six of these case studies were residential systems and five were commercial systems. Although, there was incomplete information, in each of the cases, we were able to use our model to provide values for the missing data. The highlighted values in table 19 are values taken directly from the case or values that have been calculated indirectly from the case. The medium value was chosen as a default for each missing value except for the GHG scenario and the equipment and installation costs. A "low value" of $12/ton was the default for GHG emissions and a "high value" or a 2007 cost was the default for equipment and installation costs.

The first case study that we reviewed came from the Austin Energy Report. The purposes of the Austin Energy Report were to quantify the value of distributed PV to Austin Energy, and to document the evaluation methodologies to assist Austin Energy in performing the analysis as conditions change. We completed the case study by estimating the values not quantified in the report. The values that were quantified in the report are:

- Central power generation cost
- Central power capacity cost
- T&D costs
- System losses
- Ancillary services
- Criteria pollutant emissions
- GHG emissions
- Implicit value
- Hedge value
- Ancillary services.

A fixed, 30-degree south-facing PV scenario was used in the base case. Clean Power Research has done various scenarios with different orientations and rotations. National averages or Texas region averages (when available) were used for the costs and transfers base case. As table 19 and table 20 illustrate, customer participants have a negative value while utilities/ratepayers and society have positive values. It is important to note that in this case study, the central power capacity cost is unusually low because the displaced marginal resource was a base-load gas turbine plant, compared to other studies that consider peaking-plants with limited hours of annual operation as the displaced marginal resource. Another value that is unusually low is the T&D cost benefit because load growth is occurring mostly in suburban areas with relatively low T&D costs.

Two cases were reviewed for Austin Energy – a residential system analysis and a commercial system analysis. The commercial system has a higher net value because the levelized cost of equipment and installation is lower.

The third case we reviewed was for a newly constructed residential home in Northfield, Minnesota.[73] The system consisted of 180 roof-mounted solar shingles with a combined nameplate capacity of 3.1 kW. The shingles are UniSolar 1 amorphous silicon thin-film cells, model SHR 17. This study was called "Minnesota Photovoltaic System Study: Residential Application." The actual equipment and installation costs, at 38.9 cents/kWh, were higher than our model predicted.

Table 19. Case 1 Summary Austin Energy Residential System

	PV Values	Customer/ Participant (¢/kWh)	Utility / Ratepayers (¢/kWh)	Society (¢/kWh)	Net (¢/kWh)
Benefits	Central Power Generation Cost	-	7.0	-	7.0
	Central Power Capacity Cost	-	1.1	-	1.1
	T&D Costs	-	0.1	-	0.1
	System Losses	-	0.6	-	0.6
	Ancillary Services		0.8		0.8
	Hedge Value		0.0*		0.0
	Customer Price Protection		0.8		0.8
	Criteria Pollutant Emissions				
	Greenhouse Gas Emissions	-	-	2.0**	2.0
	Implicit Value				
Costs	Equipment and Installation	(29.3)	-	-	(30.6)
	PV O&M Expenses	(0.1)	-	-	(0.1)
	Benefits Overhead	(0.2)			(0.2)
Transfer	PV Owner Electricity Bill	6.3	(6.3)	-	-
	Federal Incentives	1.6	-	(1.6)	-
	State Incentives	0.0	-	0.0	-
	Stakeholder Total	(20.9)	3.3	0.4	(17.2)

* Hedge value included in generation value ** Case looked as these values as one entity.

---

[73] *Minnesota Photovoltaic System Case Study: Residential Application.* Prepared for Iowa Department of Natural Resources. Copyright Black & Veatch (May, 2003)

Table 20. Case 2 Summary Austin Energy Commercial System

	PV Values	Customer/ Participant (¢/kWh)	Utility / Ratepayers (¢/kWh)	Society (¢/kWh)	Net (¢/kWh)
Benefits	Central Power Generation Cost	-	7.0	-	7.0
	Central Power Capacity Cost	-	1.1	-	1.1
	T&D Costs	-	0.1	-	0.1
	System Losses	-	0.6	-	0.6
	Ancillary Services		0.8		0.8
	Hedge Value		0.0*		0.0
	Customer Price Protection	0.8			0.8
	Criteria Pollutant Emissions				
	Greenhouse Gas Emissions	-	-	2.0**	2.0
	Implicit Value				
Costs	Equipment and Installation	(26.5)	-	-	(26.5)
	PV O&M Expenses	(0.1)	-	-	(0.1)
	Benefits Overhead	(0.2)			(0.2)
Transfer	PV Owner Electricity Bill	4.6	(4.6)	-	-
	Federal Incentives	7.9	-	(7.9)	-
	State Incentives	0.0	-	0.0	-
	Stakeholder Total	(13.4)	4.9	(5.9)	(14.4)

Source: Highlighted values from case study. Other values estimated from NCI model.

Table 21. Case 3 Summary New Residential Home in Northfield, MN

	PV Values	Customer/ Participant (¢/kWh)	Utility / Ratepayers (¢/kWh)	Society (¢/kWh)	Net (¢/kWh)
Benefits	Central Power Generation Cost	-	6.8	-	6.8
	Central Power Capacity Cost	-	6.0	-	6.0
	T&D Costs	-	5.1	-	5.1
	System Losses	-	0.7	-	0.7
	Ancillary Services		0.8		0.8
	Hedge Value		0.5		0.5
	Customer Price Protection	0.8	-		0.8
	Criteria Pollutant Emissions			3.6	3.6
	Greenhouse Gas Emissions			4.3	4.3
	Implicit Value			1.0	1.0
Costs	Equipment and Installation	(30.9)	-	-	(30.9)
	PV O&M Expenses	(0.1)	-	-	(0.1)
	Benefits Overhead	(0.2)			(0.2)
Transfer	PV Owner Electricity Bill	7.2	(7.2)	-	-
	Federal Incentives	1.6	-	(1.6)	-
	State Incentives	7.9	-	(7.9)	-
	Stakeholder Total	(21.8)	12.5	(0.6)	(9.9)

The fourth case we reviewed was for another new residential home that was constructed in Madison, Wisconsin, in which the cost equipment and installation costs were lower and the owner's electricity bill savings were higher.[74] The system is pole-mounted with dual-axis tracking with a nameplate capacity of 1.264 kW. It consists of Kyocera panels, an SMA inverter, and a WattSun tracker. The study was called "Wisconsin Focus on Energy Case Study: Solar Energy in the City."

Table 22. Case 4 Summary New Residential Home in Madison, WI

	PV Values	Customer/ Participant (¢/kWh)	Utility / Ratepayers (¢/kWh)	Society (¢/kWh)	Net (¢/kWh)
Benefits	Central Power Generation Cost		6.8		6.8
	Central Power Capacity Cost		6.0		6.0
	T&D Costs		5.1		5.1
	System Losses		0.7		0.7
	Ancillary Services		0.8		0.8
	Hedge Value		0.5		0.5
	Customer Price Protection	0.8			0.8
	Criteria Pollutant Emissions			2.1	2.1
	Greenhouse Gas Emissions			0.7	0.7
	Implicit Value			1.0	1.0
Costs	Equipment and Installation	(23.8)			(23.8)
	PV O&M Expenses	(0.1)			(0.1)
	Benefits Overhead	(0.2)			(0.2)
Transfer	PV Owner Electricity Bill	10.3	(10.3)		-
	Federal Incentives	1.6		(1.6)	-
	State Incentives	0.0		0.0	-
	Stakeholder Total	(11.4)	9.4	2.3	0.2

The fifth case was based on the DOE Million Roofs Initiative success stories and was for a residential home in Glenn Dale, Maryland.[75] The system consisted of roof-mounted amorphous silicon thin-film cells and had a nameplate capacity of 1.4 kW. It used BP Solarex Millennia panels and an Omnion inverter. It was indeed a success as the equipment and installation costs were low and the savings were high, yielding a relatively high net value.

---

[74] *Wisconsin Focus On Energy Case Study: Solar Energy in the City.* REN-2044-1 104 (2004)

[75] *Million Solar Roofs Initiative: Success Stories. Residential Installation, On-Grid PV System,* DOE, Glenn Dale, MD.

Table 23. Case 5 Summary Existing Residential Home in Glenn Dale, MD

	PV Values	Customer/ Participant (¢/kWh)	Utility/ Ratepayers (¢/kWh)	Society (¢/kWh)	Net (¢/kWh)
Benefits	Central Power Generation Cost	-	6.8	-	6.8
	Central Power Capacity Cost	-	6.0	-	1.1
	T&D Costs	-	5.1	-	0.1
	System Losses	-	0.7	-	0.6
	Ancillary Services		0.8		0.8
	Hedge Value		0.5		0.0
	Customer Price Protection	0.8	-		0.5
	Criteria Pollutant Emissions			2.1	2.1
	Greenhouse Gas Emissions			1.5	1.5
	Implicit Value			1.0	1.0
Costs	Equipment and Installation	(23.0)	-	-	(23.8)
	PV O&M Expenses	(0.1)	-	-	(0.1)
	Benefits Overhead	(0.2)			(0.2)
Transfer	PV Owner Electricity Bill	10.3	(10.3)	-	-
	Federal Incentives	1.6	-	(1.6)	-
	State Incentives	0.0	-	0.0	-
	Stakeholder Total	**(11.4)**	**9.4**	**2.3**	**(4.8)**

The following study was a case study on the New York Energy $Mart program.[76] The system consisted of 90 roof-mounted panels covering 900 square feet (sq ft). It had a nameplate capacity of 14.85 kW.

In this study there were only a small amount of savings yielding a low net value.

The seventh case study was based on a study by the University Massachusetts aiming to determine the effect of PV on demand charges.[77] The system consisted of 241 horizontally roof-mounted Evergreen Solar EC 110 Modules and 17 SMA SB 1 800U inverters. It had a nameplate capacity of 26.95 kW.

---

[76] *New York Energy $mart.* Darmstadt Overhead Doors. (November, 2004)

[77] *Effect of PV on Reducing Demand Charges: Case Study of a 26 kW PV System in MA*, Ujjwal Bhattacharjee and John Duffy. Energy Engineering Program, University of Massachusetts Lowell (2006)

Table 24. Case 6 Summary Existing Commercial Building in Kingston, NY

	PV Values	Customers' Participant (¢/kWh)	Utility / Ratepayers (¢/kWh)	Society (¢/kWh)	Net (¢/kWh)
Benefits	Central Power Generation Cost	-	6.8	-	6.8
	Central Power Capacity Cost	-	6.0	-	6.0
	T&D Costs	-	5.1	-	5.1
	System Losses	-	0.7	-	0.7
	Ancillary Services		0.9		0.9
	Hedge Value		0.5		0.5
	Customer Price Protection	0.8	-		0.8
	Criteria Pollutant Emissions			1.4	1.4
	Greenhouse Gas Emissions			0.4	0.4
	Implicit Value			1.0	1.0
Costs	Equipment and Installation	(31.4)	-	-	(31.4)
	PV O&M Expenses	(0.1)	-	-	(0.1)
	Benefits Overhead	(0.2)			(0.2)
Transfer	PV Owner Electricity Bill	1.7	(1.7)	-	-
	Federal Incentives	7.9	-	(7.9)	-
	State Incentives	17.8	-	(17.8)	•
	Stakeholder Total	**(3.5)**	**18.0**	**(23.0)**	**(8.6)**

Table 25. Case 7 Summary Existing Commercial Building in Cambridge, MA

	PV Values	Customers' Participant (¢/kWh)	Utility / Ratepayers (¢/kWh)	Society (¢/kWh)	Net (¢/kWh)
Benefits	Central Power Generation Cost	-	6.8	-	6.8
	Central Power Capacity Cost	-	6.0	-	6.0
	T&D Costs	-	5.1	-	5.1
	System Losses	-	0.7	-	0.7
	Ancillary Services		0.9		0.9
	Hedge Value		0.5		0.5
	Customer Price Protection	0.8	-		0.8
	Criteria Pollutant Emissions			1.4	1.4
	Greenhouse Gas Emissions			0.3	0.3
	Implicit Value			1.0	1.0
Costs	Equipment and Installation	(52.1)	-	-	(31.4)
	PV O&M Expenses	(0.1)	-	-	(0.1)
	Benefits Overhead	(0.2)			(0.2)
Transfer	PV Owner Electricity Bill	1.1	(1.1)	-	-
	Federal Incentives	7.9	-	(7.9)	-
	State Incentives	7.9	-	(7.9)	•
	Stakeholder Total	**(34.7)**	**19.0**	**(13.2)**	**(29.4)**

In this chapter there were very large equipment and installation costs and low electricity bill savings, again yielding a very low net value.

The next case was a study by the San Diego Regional Energy Office.[78] The system consisted of 9,700 roof-mounted panels with a total nameplate capacity of 970 kW. It was installed on the Del Mar Fairgrounds of San Diego County.

Table 26. Case 8 Summary, Existing Municipal in Del Mar, CA

	PV Values	Customer/ Participant (¢/kWh)	Utility/ Ratepayers (¢/kWh)	Society (¢/kWh)	Net (¢/kWh)
Benefits	Central Power Generation Cost	-	6.8	-	6.8
	Central Power Capacity Cost	-	6.0	-	6.0
	T&D Costs	-	5.1	-	5.1
	System Losses	-	0.7	-	0.7
	Ancillary Services		0.8		0.8
	Hedge Value		0.5		0.5
	Customer Price Protection	0.8	-		0.8
	Criteria Pollutant Emissions			1.4	1.4
	Greenhouse Gas Emissions			2.9	2.9
	Implicit Value			1.0	1.0
Costs	Equipment and Installation	(17.8)	-	-	(17.8)
	PV O&M Expenses	(0.1)	-	-	(0.1)
	Benefits Overhead	(0.2)			(0.2)
Transfer	PV Owner Electricity Bill	15.2	(15.2)	-	-
	Federal Incentives	7.9	-	(7.9)	-
	State Incentives	11.3	-	(11.3)	-
	Stakeholder Total	17.2	4.4	(14.0)	7.6

This retrofit had very high value because of the low equipment and installation costs and high electricity bill savings.

Case nine was based on a study by the Solar Washington Newsletter called "Solar Power Case Study: An Old Idea with New Economics."[79] The system consisted of eight roof-mounted Sanyo panels totaling 104 sq ft. It had a nameplate capacity of 1.52 kW.

Table 27. Case 9 Summary Existing Residential Home in Bellingham, WA

---

[78] San Diego Regional Energy Office. Del Mar Fairgrounds SelfGen Case Study

[79] *Solar Power Case Study: An Old Idea With New Economics.* John Watts. Solar Washington Newsletter. (Fall, 2006)

PV Values		Customer/ Participant (¢/kWh)	Utility/ Ratepayers (¢/kWh)	Society (¢/kWh)	Net (¢/kWh)
Benefits	Central Power Generation Cost	-	6.8	-	6.8
	Central Power Capacity Cost	-	6.0	-	6.0
	T&D Costs	-	5.1	-	5.1
	System Losses	-	0.7	-	0.7
	Ancillary Services		0.8		0.8
	Hedge Value		0.5		0.5
	Customer Price Protection	0.8	-		0.8
	Criteria Pollutant Emissions			0.7	0.7
	Greenhouse Gas Emissions			0.1	0.1
	Implicit Value			1.0	1.0
Costs	Equipment and Installation	(33.3)	-	-	(33.3)
	PV O&M Expenses	(0.1)	-	-	(0.1)
	Benefits Overhead	(0.2)			(0.2)
Transfer	PV Owner Electricity Bill	7.7	(7.7)	-	-
	Federal Incentives	1.6	-	(1.6)	-
	State Incentives	2.2	-	(2.2)	-
	Stakeholder Total	(21.3)	11.9	(2.0)	(11.4)

Again, high equipment and installation combined with average bill savings resulted in a low net value.

Cases ten and eleven were based on a study by the Center for Sustainable Energy for Million Solar Roofs Initiative in January 2007, called "New York City's Solar Energy Future Part II: Solar Energy Policies and Barriers in New York City."[80] Low installation costs and high incentives created a net positive value.

Table 28. Case 10 Summary Existing Residential Home in New York, NY

---

[80] *New York City's Solar Energy Future Part II: Solar Energy Policies and Barriers in New York City.* Prepared by the Center for Sustainable Energy at Bronx Community College for The City University of New York's Million Solar Roofs Initiative. Appendix VI PV Planner Assumptions for all Sectors. Jan. 2007

	PV Values	Customer/ Participant (¢/kWh)	Utility / Ratepayers (¢/kWh)	Society (¢/kWh)	Net (¢/kWh)
Benefits	Central Power Generation Cost	-	6.8	-	6.8
	Central Power Capacity Cost	-	6.0	-	6.0
	T&D Costs	-	5.1	-	5.1
	System Losses	-	0.7	-	0.7
	Ancillary Services		0.8		0.8
	Hedge Value		0.5		0.5
	Customer Price Protection	0.8	-		0.8
	Criteria Pollutant Emissions			1.4	1.4
	Greenhouse Gas Emissions			0.3	0.3
	Implicit Value			1.8	1.8
Costs	Equipment and Installation	(14.7)	-	-	(14.7)
	PV O&M Expenses	(1.3)	-	-	(1.3)
	Benefits Overhead	(0.2)			(0.2)
Transfer	PV Owner Electricity Bill	26.4	(26.4)	-	-
	Federal Incentives	1.6	-	(1.6)	-
	State Incentives	16.8	-	(16.8)	-
	Stakeholder Total	**29.3**	**(6.7)**	**(15.7)**	**6.9**

Table 29. Case 11 Summary Existing Commercial Building in New York, NY

	PV Values	Customer/ Participant (¢/kWh)	Utility / Ratepayers (¢/kWh)	Society (¢/kWh)	Net (¢/kWh)
Benefits	Central Power Generation Cost	-	6.8	-	6.8
	Central Power Capacity Cost	-	6.0	-	6.0
	T&D Costs	-	5.1	-	5.1
	System Losses	-	0.7	-	0.7
	Ancillary Services		0.8		0.8
	Hedge Value		0.5		0.5
	Customer Price Protection	0.8	-		0.8
	Criteria Pollutant Emissions			1.4	1.4
	Greenhouse Gas Emissions			0.3	0.3
	Implicit Value			1.8	1.8
Costs	Equipment and Installation	(14.7)	-	-	(14.7)
	PV O&M Expenses	(1.3)	-	-	(1.3)
	Benefits Overhead	(0.2)			(0.2)
Transfer	PV Owner Electricity Bill	15.7	(15.7)	-	-
	Federal Incentives	7.9	-	(7.9)	-
	State Incentives	17.8	-	(17.8)	-
	Stakeholder Total	**29.3**	**(6.3)**	**(15.7)**	**7.4**

4.2. Scenarios

In addition to the case studies, we created five scenarios that demonstrate the model capability. The scenarios are a PV system with storage, a low installation and equipment cost scenario, a low installation and equipment cost scenario with no incentives, a $30/ton GHG scenario, and a $50/ton GHG scenario.

Base Scenario

The base scenario was a retrofit of a residential home in San Diego, California. Below is the base scenario table of values.

Table 30. Base Scenario

	PV Values	Customer/ Participant (¢/kWh)	Utility/ Ratepayer (¢/kWh)	Society (¢/kWh)	Net (¢/kWh)
Benefits	Central Power Generation Cost	*	6.8	*	6.8
	Central Power Capacity Cost	-	6.0	-	6.0
	T&D Costs	-	5.1	-	5.1
	System Losses	-	0.7	-	0.7
	Ancillary Services		0.8		0.8
	Hedge Value		0.5		0.5
	Customer Price Protection	0.8	-		0.8
	Criteria Pollutant Emissions			1.4	1.4
	Greenhouse Gas Emissions			0.4	0.4
	Implicit Value			1.0	1.0
Costs	Equipment and Installation	(29.3)	-	-	(29.3)
	PV O&M Expenses	(0.1)	*	*	(0.1)
	Benefits Overhead	(0.2)			(0.2)
Transfer	PV Owner Electricity Bill	14.0	(14.0)	-	-
	Federal Incentives	2.6	-	(2.6)	-
	State Incentives	12.0	*	(12.0)	*
	Stakeholder Total	**0.7**	**5.7**	**(12.9)**	**(6.4)**

Scenario 1 - Storage

The first scenario is a PV system with storage. Values were obtained from the Storage and Controls Report from DOE's Office of Energy Efficiency and Renewable Energy (EERE) and NREL.[81] In this chapter, the additional value of outage protection provided by the storage was $245/year for a residential system. The cost of the battery (3 kWh) was $900 and the lifetime of the battery was 7 years. Thus, the additional cost was 1.40 cents/kWh and the value of reliability was 2.7 cents/kWh. Another potential scenario could involve customers using more expensive storage systems for energy arbitrage opportunities in addition to back-up power.

Table 31. Scenario 1 - Storage

---

[81] Hoff, T.E., Perez, R., Braun, Margolis, R M., Maximizing the Value of Customer-Sited PV Systems Using Storage and Controls, NREL (2005)

	PV Values	Customer/ Participant (¢/kWh)	Utility/ Ratepayers (¢/kWh)	Society (¢/kWh)	Net (¢/kWh)
Benefits	Central Power Generation Cost	-	6.5	-	6.5
	Central Power Capacity Cost	-	6.0	-	6.0
	T&D Costs	-	5.1	-	5.1
	System Losses	-	2.4	-	2.4
	Ancillary Services	-	0.8	-	0.8
	Hedge Value	-	0.5	-	0.5
	Customer Price Protection	0.8	-	-	0.8
	Customer Reliability	2.7	-	-	2.7
	Criteria Pollutant Emissions	-	-	1.4	1.4
	Greenhouse Gas Emissions	-	-	0.4	0.4
	Implicit Value	-	-	1.0	1.0
Costs	Equipment and Installation	(30.7)	-	-	(30.7)
	PV O&M Expenses	(0.1)	-	-	(0.1)
	Benefits Overhead	(0.2)	-	-	(0.2)
Transfer	PV Owner Electricity Bill	14.0	(14.0)		-
	Federal Incentives	2.6	-	(2.6)	-
	State Incentives	12.0	-	(12.0)	-
	Stakeholder Total	**2.0**	**7.1**	**(12.8)**	**(3.7)**

Scenario 2-Low Cost

The second scenario is a low installation and equipment cost scenario. The values for this scenario were obtained from our model. The assumptions for this scenario included projecting the costs out to 2015, which decrease annually at 4% to 6% per year.

Table 32. Scenario 2- Low Cost

	PV Values	Customer/ Participant (¢/kWh)	Utility/ Ratepayers (¢/kWh)	Society (¢/kWh)	Net (¢/kWh)
Benefits	Central Power Generation Cost	-	6.8	-	6.8
	Central Power Capacity Cost	-	10.0	-	10.8
	T&D Costs	-	10.0	-	10.0
	System Losses	-	0.7	-	0.7
	Ancillary Services	-	0.8	-	0.8
	Hedge Value	-	0.5	-	0.5
	Customer Price Protection	0.8	-	-	0.8
	Criteria Pollutant Emissions	-	-	1.4	1.4
	Greenhouse Gas Emissions	-	-	0.4	0.4
	Implicit Value	-	-	1.0	1.0
Costs	Equipment and Installation	(19.0)	-	-	(19.0)
	PV O&M Expenses	(0.1)	-	-	(0.1)
	Benefits Overhead	(0.2)	-	-	(0.2)
Transfer	PV Owner Electricity Bill	14.0	(14.0)	-	-
	Federal Incentives	2.6	-	(2.6)	-
	State Incentives	12.0	-	(12.8)	-
	Stakeholder Total	**11**	**5.7**	**(12.8)**	**3.9**

This created a net positive value and a positive value for the customer/participant. The third scenario is a low cost installation and equipment cost scenario with no incentives.

Scenario 3 − Low Cost and No Incentives
We used the same assumptions as in the previous scenario and did not include any state or federal incentives in the value analysis.

Table 33. Scenario 3. Low Cost and No Incentives

	PV Values	Customers' Participant (¢/kWh)	Utility/ Ratepayers (¢/kWh)	Society (¢/kWh)	Net (¢/kWh)
Benefits	Central Power Generation Cost	-	6.8	-	6.8
	Central Power Capacity Cost	-	10.8	-	10.8
	T&D Costs	-	10.0	-	10.0
	System Losses	-	0.7	-	0.7
	Ancillary Services	-	0.8	-	0.8
	Hedge Value	-	0.5	-	0.5
	Customer Price Protection	0.8	-	-	0.8
	Criteria Pollutant Emissions	-	-	1.4	1.4
	Greenhouse Gas Emissions	-	-	0.4	0.4
	Implicit Value	-	-	1.0	1.0
Costs	Equipment and Installation	(19.0)	-	-	(19.0)
	PV O&M Expenses	(0.1)	-	-	(0.1)
	Benefits Overhead	(0.2)	-	-	(0.2)
Transfer	PV Owner Electricity Bill	14.0	(14.0)	-	-
	Federal Incentives	-	-	-	-
	State Incentives	-	-	-	-
	Stakeholder Total	(4.5)	(5.7)	2.7	3.9

The overall net value did not change from scenario 2 and remained positive; however, the benefit to the customer participant decreased and the benefit to society increased.

Scenario 4 - GHG Emissions at $30/ton
The fourth scenario was our first GHG scenario. We used a value of $30/ton of CO_2 for the value of GHG − a value we thought would have a significant impact on carbon emissions in the United States.

Table 34. Scenario 4 Greenhouse Gas Emissions at $30/ton

	PV Values	Customer/Participant (¢/kWh)	Utility/Ratepayers (¢/kWh)	Society (¢/kWh)	Net (¢/kWh)
Benefits	Central Power Generation Cost	-	6.8	-	6.8
	Central Power Capacity Cost	-	10.9	-	10.9
	T&D Costs	-	10.0	-	10.0
	System Losses	-	0.7	-	0.7
	Ancillary Services	-	0.8	-	0.8
	Hedge Value	-	0.5	-	0.5
	Customer Price Protection	0.8	-	-	0.8
	Criteria Pollutant Emissions	-	-	1.4	1.4
	Greenhouse Gas Emissions	-	-	0.9	0.9
	Implicit Value	-	-	1.8	1.8
Costs	Equipment and Installation	(29.3)	-	-	(29.3)
	PV O&M Expenses	(0.1)	-	-	(0.1)
	Benefits Overhead	(0.2)	-	-	(0.2)
Transfer	PV Owner Electricity Bill	14.0	(14.0)	-	-
	Federal Incentives	2.6	-	(2.6)	-
	State Incentives	12.0	-	(12.0)	-
	Stakeholder Total	0.7	5.7	(12.2)	(5.9)

This assumption increased the value of GHG emissions to 0.9 cents/kWh; however, benefits to society were still negative overall.

Scenario 5 − GHG Emissions at $50 per ton

Our last scenario was our second GHG scenario. We used a value of $50/ton of carbon dioxide; a value we thought the market had the potential to reach in the long term. Table 35 highlights the results of this analysis.

Table 35. Scenario 5 Greenhouse Gas Emissions at $50/ton

	PV Values	Customer/Participant (¢/kWh)	Utility/Ratepayers (¢/kWh)	Society (¢/kWh)	Net (¢/kWh)
Benefits	Central Power Generation Cost	-	6.8	-	6.8
	Central Power Capacity Cost	-	10.9	-	10.9
	T&D Costs	-	10.0	-	10.0
	System Losses	-	0.7	-	0.7
	Ancillary Services	-	0.8	-	0.8
	Hedge Value	-	0.5	-	0.5
	Customer Price Protection	0.8	-	-	0.8
	Criteria Pollutant Emissions	-	-	1.4	1.4
	Greenhouse Gas Emissions	-	-	1.5	1.5
	Implicit Value	-	-	1.0	1.0
Costs	Equipment and Installation	(29.3)	-	-	(29.3)
	PV O&M Expenses	(0.1)	-	-	(0.1)
	Benefits Overhead	(0.2)	-	-	(0.2)
Transfer	PV Owner Electricity Bill	14.0	(14.0)	-	-
	Federal Incentives	2.6	-	(2.6)	-
	State Incentives	12.0	-	(12.0)	-
	Stakeholder Total	0.7	5.7	(11.6)	(5.2)

This assumption increased the greenhouse gas emissions value to 1.5 cents/kWh.

5.0. PV VALUE GAP ANALYSIS

5.1. Increasing Benefits from Key PV Values

We identified gaps between what is needed and what is available so as to make recommendations for improving the value from PV systems. Opportunities to increase the benefits or reduce the costs were focused on PV values with the highest magnitudes. Several strategies were identified for each of the key PV values. The paragraphs below discuss the key PV values and potential ways to improve the benefits or reduce the costs.

Central Power Generation Cost
Natural gas-fueled power plants are the marginal generation resource in many regions of the United States. As a result, natural gas prices in those regions and the marginal resource heat rate (i.e., the amount of gas consumed to generate a kilowatt-hour) are two key drivers of this value. Given these drivers, a PV system in a region with high gas prices that often displaces electricity from an inefficient peaking power plant will have a much higher benefit than a PV system in a region with low gas prices that displaces mainly off-peak power from an efficient baseload power plant. A strategy to improve the benefit from this value is to increase production from the PV system by optimizing orientation (i.e., latitude and tilt) and using tracking systems.

Central Power Capacity Cost
The key driver for this value is the coincidence of peak demand with system output. Another key driver is the type of generation asset displaced. Peaking plants typically have lower capital costs than baseload plants. However, a peaking plant that runs a limited number of hours per year will have a higher capital cost per kilowatt-hour than a baseload plant. Given these drivers, a system that produces a high share of its output during on- peak hours and displaces a peaking plant will have a higher benefit. Various strategies to increase production during peak demand periods and increase the benefit from this value include: using thin film technology that has a more consistent output during the day, integrating energy storage into the PV system, and integrating load management applications with the PV system controls.

T&D Cost
While this value has significant potential, it has been difficult to capture. This value depends on the location of the PV system as well as the output during the T&D system's peak period. Locations with congested transmission and/or distribution systems typically require expensive upgrades that could be deferred where PV systems are installed to reduce congestion. Although this value includes both T&D, there are cases that are specific to one or the other. For example, PV can be installed in an area that reduces the transmission peak, but is in a distribution area with excess capacity, providing limited value to the distribution system. However, some of the most congested areas have

network distribution systems in which interconnection standards currently prohibit or severely limit interconnections. And in non-network distribution systems, the deferral depends on the production of the PV system during the peak of the specific distribution area, which varies across the distribution system and can be a different peak period than the regional generation peak.

Many distribution planners will also want "physical assurance" (i.e., guarantees that the load the PV is serving is permanently displaced). Another barrier for this value is the potential for circuit overload following an outage or a recloser operation. Current interconnection standards (e.g., IEEE 1547) prohibit PV systems from riding through outages, leaving the T&D system to support the loads that would otherwise be served by the PV systems. For a local utility to defer T&D upgrades and capture the benefits of this value, it will need to assure that the PV system will effectively eliminate a certain load during its peak congestion period and that, if the PV system trips, the customer's connected load will be reduced by an amount equal to the PV output. PV integrated with load management systems would address these concerns and allow PV systems to capture the T&D cost benefits. Other strategies to increase the benefits captured from this value are to improve the ability to install PV systems in congested areas with limited roof space, and to firm PV output with storage and/or other demand side actions.

Greenhouse Gas and Criteria Pollutant Emissions

This value is driven by two key factors: the amount of emissions displaced by the PV system and the value of the displaced emissions. PV systems have no emissions and displace all the emissions associated with the marginal central generation resource. In most cases, the marginal resource will be a gas-fueled central generation plant. The higher the heat rate, the higher the displaced emissions. Coincidence of peak demand with PV system output also plays a role in this value because peaking plants tend to have a higher heat rate than baseload plants, producing higher emissions.

As for valuating the displaced emissions, there are various mechanisms currently in place: assessing the public health impact, employing regional air quality district emissions permit trading systems, employing regional renewable energy credit trading systems, assessing the penalties of failing to meet emission standards, and projecting the cost of achieving target emissions reductions. The variety of valuation mechanisms has created a wide range of economic benefits for this value. Moreover, the climate and energy context is evolving quickly and producing an upward trend in the future economic value of the emissions reduction. A strategy to increase the benefits from this value is to support the development and adoption of a uniform valuation standard across the country. A secondary strategy is to increase the amount of displaced emissions by aligning the PV system production with peak demand periods, when dirtier peaking generation plants are the marginal resource.

Implicit Value of PV

This value is driven by customers' willingness to pay a premium price for electricity from a PV system. For some commercial customers, this value could come from demonstrating to their key stakeholders (customers, investors, employees, and regulators) that the organization is environmentally friendly. For some residential customers, this value could come from a desire to reduce their environmental impact, create an image of being environmentally friendly, and/or create an image of being an early adopter of emerging technologies. The magnitude of this value across market segments is still unclear. Furthermore, this value may change over time as PV penetrates the market. The implicit value may decline as PV becomes more common. Or conversely, PV may become a "must-have" product for some sectors of the economy. Understanding what creates this value and how it will change over time will be critical to the success of PV.

Equipment and Installation Cost

This value is driven by three key factors: system size, location, and projected long-term costs. Large systems have a lower cost per output unit than smaller systems because some PV system costs (e.g., design, engineering, transportation, installation, permitting, and incentive request) are mostly fixed. The location is also a factor because labor rates in some regions are more expensive than others, driving up the labor-intensive costs (e.g., design, engineering, and installation).

As the industry continues to grow and mature, economies of scale and learning curves across the supply chain are expected to reduce overall system costs. However, it is still unclear exactly how much costs will come down, and projections have significant variance. A strategy to reduce the cost from this value is to continue to promote incentives and remove regulatory and market barriers that will help the industry grow and achieve the expected economies of scale. Another strategy is to help capture and disseminate operational best practices from Europe (Germany) and Asia (Japan) across the supply chain to help increase production and decrease costs due to economies of scale and the learning curve effect.

5.2. Integrating Storage to PV Systems

Several reports sponsored by EERE and NREL have examined the benefits of integrating storage with PV.[82][83] Both the outage protection benefits and energy arbitrage opportunities were analyzed. The report, *Maximizing the Value of Customer-Sited PV Systems Using Storage and Controls*, looked at both a residential and commercial case and found that savings due to outage protection were \$245/year and \$25,000/year respectively. The value was derived from an estimated yearly cost of outage-related disturbances prorated to the relative size of the considered PV system. Load management, outage prevention, and outage recovery were the three values of onsite storage that were defined in this report.

Another report, *Increasing the Value of Customer-Owned PV Systems Using Batteries*, found that PV storage can have value to both the consumer and utility. This report focused on determining the savings using storage without PV and with PV, and found storage savings to be significant. Cost savings due to a reduction in storage capacity increased the NPV to

---

[82] Hoff, T.E., Perez, R., Braun, Margolis, R M., Increasing the Value of Customer-Owned PV Systems Using Batteries, NREL (November 9 2004)

[83] Hoff, T.E., Perez, R., Braun, Margolis, R M., Maximizing the Value of Customer-Sited PV Systems Using Storage and Controls, NREL (2005)

$30,000 (from $5,000). Obtaining outage cost and associated outage probabilities was difficult to obtain.

While much of the previous research suggests that integrating energy storage equipment into PV systems can improve the value provided by PV systems, more quantification of the benefits for the consumer during an outage or for the utility to control consumer-owned battery systems in an emergency are areas for future research. Current energy storage technology is still relatively expensive and inefficient. However, R&D is underway that aims to develop new energy storage technologies that would have significant cost and efficiency improvements over today's technologies. Feasibility studies analyzing the economics of switching from firm to non-firm rates is another area for further study.[84] In the paragraphs below, we discuss specific examples of opportunities for storage systems to increase the benefits of PV systems.

Optimizing Output Profile during the Day

Short-term (i.e., hours) storage systems can store energy produced by the PV system during the off-peak period of the day and provide energy and capacity during on-peak periods. By adjusting the output of the PV system to coincide with the peak demand period, the storage system is increasing the benefits along several key values:

- Central power generation cost. Increases the displacement of less efficient peaking plants that have higher operating costs.
- Central power capacity cost. Firms the PV to ensure peak central generation capacity is displaced.
- T&D costs. Provides greater assurance that the PV can provide T&D capacity during peak demand periods on the T&D system to increase the deferral of T&D upgrades.
- System losses. Increases the avoided losses because they are higher during peak periods.
- GHG and criteria pollutant emissions. Increases the displacement of lower efficiency peaking plants that have higher criteria pollutants.

Increasing Reliability of System Output

Long-term (i.e., days) energy storage can store energy from the PV system on days with good insolation and release it on days with poor insolation. The storage system is therefore increasing the likelihood that the PV system will be able to consistently offset a certain level of demand. However, it is very expensive to have "days" of storage and most of the benefit can be obtained with small amounts of storage and load control.[85] Customers will also have reliable back-up power systems that are based on renewable energy, rather than diesel generation. The values with increased benefits from this application are:

---

[84] Hoff, T.E., Perez, R., Margolis, R., Increasing the Value of Customer-Owned PV Systems Using Batteries (November 9, 2004)

[85] Hoff, T.E., Perez, R., Braun, Margolis, R M., Maximizing the Value of Customer-Sited PV Systems Using Storage and Controls, NREL (2005)

- Central power capacity cost. Increases the confidence that PV systems in a certain region will provide a certain level of power on peak demand days and increase the displaced peaking capacity.
- T&D costs. Increases the confidence that the PV system will provide a certain amount of power on peak demand days and increase the opportunities to defer T&D investments.
- System resiliency and customer reliability. Increases the confidence that PV systems can provide power during outages. Provides utilities with a resource to supplement the ability of PV to relieve stress on their T&D systems while supporting customers' critical loads and keeping businesses and residences running.[86]

5.3. Integrating Demand Response with PV Systems

A less expensive alternative to energy storage is demand response (also called demand side management or load control applications). By integrating PV systems with demand response, PV systems can more effectively reduce peak demand and the associated central power capacity and T&D costs. This may be one of the best ways to increase market penetration of PV. When demand is at its highest, relatively expensive PV should be running to capture the market share. Thus, it may serve PV in the long run to assess how it could become fully integrated with demand side management both locally (at point of use) or across the power grid under the potential control of the utility. It could potentially be used to alleviate stress on the grid.

When the output of the PV system is coincident with peak load, demand response can provide assurances to system operators and allow them to rely on the integrated photovoltaic/demand response (PV/DR) system to provide resource adequacy, peak capacity, and peak energy. PV also brings value to demand response. With an integrated PV/DR system the utility would rely less on demand response and the customer would be subject to fewer demand response events.

NREL and EERE reports examined demand response with PV systems.[87][88] In the report, *Maximizing the Value of Customer-Sited PV Systems Using Storage and Controls*, cumulative cash flows for PV, emergency storage, PV and local load management storage, and PV and emergency storage were compared. In the San Jose commercial case, PV and load control storage had an NPV of $212,000 while PV alone had an NPV of $173,000. In the commercial Long Island case, PV and load control storage had an NPV of $21,500 while PV alone had an NPV of $7,500. This study concluded that only a small addition of storage for local load control is beneficial for customer-sited PV. In the Report, *Maximizing PV Peak Shaving with Solar Load Control: Validation of a Web- Based Economic Evaluation Tool*, it was found that solar load control can substantially enhance PV

---

[86] Hoff, T.E., Perez, R., Braun, Margolis, R M., Maximizing the Value of Customer-Sited PV Systems Using Storage and Controls, NREL (2005)

[87] Ibid.

[88] Perez, R., Hoff, T.E., Herig, C., Shah, J., Maximizing PV Peak Shaving with Solar Load Control: Validation of a Web-Based Economic Evaluation Tool (May, 2003)

load reduction. The solar load control NPV ranged from $10,000 to $100,000 depending on the size of the building.

6.0. RECOMMENDATION FOR FUTURE RESEARCH

A wealth of previous research has been performed to evaluate the value of PV. However, specific research still needs to be completed to fill in knowledge gaps. We identified opportunities for DOE and NREL to sponsor additional R&D efforts to improve the quantification of PV values and/or improve the value captured from PV systems. These research efforts will help improve the value of grid connected PV systems, reach economic parity with traditional central generation options, and increase their market penetration. We grouped our recommendations into short-, medium- and long-term focus areas for DOE R&D.

SHORT-TERM R&D RECOMMENDATIONS (1-3 YEARS)

Over the next couple of years, DOE should promote a standard framework and develop tools easily available to industry to assess the value of PV systems. With a few exceptions, our research shows that most of the high magnitude PV values had well-established methodologies to quantify the financial impacts, while some of the PV values with lower magnitude PV values do not have generally accepted methods. Although many of these high magnitude values have well established methodologies, there is still disagreement about input assumption, and this is a critical issue.

DOE should look at existing programs that assess the value of PV and could support the next R&D steps. These include: the Massachusetts Technology Collaborative (MTC), Sacramento Municipal Utility District (SMUD), and Austin Energy. For example, the MTC is undertaking a pilot program to place a large amount of PV within a congested area of the grid to test the actual values of PV. DOE might consider offering to review the results, leverage some dollars to support the analysis of results, and take next steps in funding follow-on R&D. DOE can also support additional analyses of PV values within utility distribution networks, which the California Energy Commission (CEC) is currently funding.

In addition, DOE should take a leadership role in the development of a standard approach to value GHG and criteria pollutant emissions, including the associated health benefits and intrinsic value of reducing emissions. These are two areas of contention among existing studies that quantify the value of PV. Some studies had a conservative view of benefits while others had high expectations for the future.

Another high-priority, short-term R&D opportunity for DOE is to fund a robust effort to quantify the costs and benefits associated with integrating current and emerging energy storage systems and demand response applications with PV systems. This effort should go beyond existing efforts and assess both costs and benefits in an approach that involved key stakeholders, including utilities and customers.

MID-TERM R&D RECOMMENDATIONS (3-5 YEARS)

Over the mid-term, DOE should collaborate with utilities in the development and deployment of new technologies and operating practices to increase the value captured by utilities and ratepayers from PV systems. These research efforts will help improve the value of grid connected PV systems, reach economic parity with traditional central generation options, and increase their market penetration. Specific R&D opportunities by PV value are presented in table 36.

LONG-TERM R&D RECOMMENDATIONS (5-10 YEARS)

Over the long term, DOE should take a leadership role in establishing frameworks for long-term policies, regulations, and incentives that will reduce the risks and uncertainty currently limiting investment in PV markets. Solving the technical and analytical challenges over the short- and mid-term will certainly help advance the adoption of PV systems. But to achieve a broad market penetration, appropriate policies, regulations, and incentives will need to be in place. DOE will need to support the development of these mechanisms.

Table 36. Recommended R&D to Increase the Value of Grid Interconnected PV

PV Value	R&D Opportunities
Central Power Capacity Cost	• Assess opportunity to store energy from PV systems on a distributed storage system located on the distribution feeder • Generally improve the cost and performance of energy storage systems • Understand how demand response can help PV systems displace central capacity • Investigate potential improvements to optics in glass of PV modules to enhance performance • Assess the potential for stranded generation assets for new fossil-fuel power plant investments due to significant changes in fuel costs and/or environmental regulations. Assess the levelized cost of energy from such plants should their economic life be reduced from 30 to 20 or even to 10 years
Central Power Generation Cost	• As PV reaches significant penetration levels (e.g., >5%), develop guidelines and tools for system planners to predict performance of PV systems (e.g., a day ahead based on weather information) and integrate loading forecasting for wholesale power markets • Assess actual performance of installed PV systems against predicted performance

Table 36. (Continued).

PV Value	R&D Opportunities
T&D Costs	• Assess the technical limit for PV penetration on a single distribution circuit (e.g., currently a 15% compromise agreement in Massachusetts) and define strategies to increase the penetration limit • Develop the capability to more easily locate PV systems into the "built environment" of congested areas with limited roof space (e.g., PV integrated with roads, bridges and/or rail lines) • Develop cost-effective engineering solutions to interconnect PV systems to network distribution systems • Develop cost-effective storage systems • Explore potential to bundle PV systems with batteries in electric/hybrid vehicles • Develop cost-effective monitoring, control and validation for performance and physical assurance • Develop cost-effective engineering solutions to allow PV systems to ride through system outages • Assess opportunity to integrate demand response to PV systems to improve T&D benefits • Assess the technical and economic feasibility of PV-ready distribution systems • Develop tools for utilities to incorporate PV system output uncertainty into T&D planning and capacity deferment processes • Develop easy to use tools for regulators to quantify T&D values of PV systems in their regions
Greenhouse Gas and Criteria Pollutant Emissions	• Launch a consensus based effort to standardize emissions valuation method. While there is some understanding of how to quantify the greenhouse gas emissions displaced by PV systems, there are many approaches being used to value the displaced emissions. This issue goes beyond PV to all greenhouse gas emissions reduction applications. Even though there is much research underway, it is likely that more research will be required to develop a standard valuation approach that is widely accepted and used across the country.
Implicit Value of PV	• Use surveys/focus groups to understand motivations within each stakeholder group (e.g., consumers, developers, utilities) behind PV projects that went ahead even with a negative net present value • Develop approaches to monetize the value of enhancing the corporate brand • Understand how the implicit value of PV will change over time; how long
System Resiliency	• Collect data from utilities or on a region-wide basis regarding number of hours of power outage and assess which outage hours could have been avoided by the use of PV with distributed energy storage
Customer Reliability	• Assess the value of backup power for critical loads that can be provided by PV with distributed energy storage for residential and commercial

PV Value	R&D Opportunities
Customer Electric Price Protection	• Develop an accepted methodology to value the certainty of fuel prices • Develop improved models to forecast long-term electricity prices • Use surveys/focus groups to understand if and how much price protection is a motivator for PV projects that went ahead even with a negative net present value • Evaluate current mechanisms used by consumers for energy price hedging
Equipment and Installation Cost	• Compare technology requirements in Europe, Japan and the U.S. and assess the impact of codes on PV system design and costs • Capture and disseminate operational best practices from Europe and Asia across the supply chain that will accelerate the learning curve and reduce costs • Assess potential for distributing regulated low voltage, PV generated dc power to electronic devices in residential and commercial buildings and the savings from eliminating multiple ac-to-dc power supplies and
Benefits Overhead	• Understand what data and infrastructure (e.g., meters, data management, remote controlling and central monitoring) is required, and what it will take for utilities to rely on PV as a resource • Understand incentive program management costs (e.g., incentive administration, cost to government, cost to PUCs, cost to installers, cost to customers) and identify ways to reduce these costs (e.g., incentive program standardization across states) • Develop tools for industry to quickly and easily perform preliminary analysis on the value of specific PV installation projects

CONCLUSION

This chapter provides a series of important conclusions regarding the analysis of PV value:

- Previous efforts to quantify PV values focus on a single stakeholder, typically the participant customer or the utility. When the values across key stakeholders are aggregated, several significant values simply become transfers from one stakeholder to another without creating a net benefit or cost (e.g., PV owner savings on electricity bill, federal incentives, and state incentives).
- While 19 PV values were identified, only 6 have significant benefits (central power generation cost, central power capacity cost, T&D costs, GHG emissions, criteria pollutant emissions, and implicit value of PV) and one has significant costs (equipment and installation cost).
- The net value of PV systems varies greatly, driven primarily by the location of the system and the output profile (time of day/season).
- Several PV values require additional R&D to establish a standardized quantification methodology (implicit value of PV, system resiliency, fuel diversity, market price impacts/elasticity, and customer reliability).

- While much of the previous research suggests that integrating energy storage equipment into PV systems can improve the value provided by PV systems, more quantification of the costs and benefits is required.
- There are still many opportunities to increase the benefits of PV systems through R&D.

It is recommended that NREL and DOE enhance their efforts to fund R&D that will increase the magnitude and clarity of value from grid connected PV systems. More specifically:

Over the short-term

- Promote a standard framework and develop tools easily available to industry to assess the value of PV systems.
- Take a leadership role in the development of a standard approach to value GHG and criteria pollutant emissions.
- Quantify the costs and benefits associated with integrating current and emerging energy storage systems and demand response applications with PV systems.

Over the mid-term

- Collaborate with utilities in the development and deployment of new technologies and operating practices to increase the value captured by utilities and ratepayers from PV systems.

Over the long-term

- Take a leadership role in establishing frameworks for long-term policies, regulations, and incentives that will reduce the risks and uncertainty currently limiting investment in PV markets.

REFERENCES

ABT Associates; ICF Consulting, (October 2000). *The Particulate-Related Health Benefits of Reducing Power Plant Emissions* http://www.cleartheair.org/fact/mortality/mortalityabt.pdf

Americans for Solar Power (April 13, 2005). *Build-Up of PV Value in California* http://www.mtpc.org/renewableenergy/public_policy/DG/resources/2005-04-CA-PVValue-Links-R04-03-017.pdf ("ASPv Report")

Bhattacharjee, U.; Duffy, J. (2006). "Effect of PV on Reducing Demand Charges: Case Study of a 26 kW PV System in MA" Energy Engineering Program, University of Massachusetts, Lowell, MA http://www.solar2006.org/presentations/tech_sessions/t40-a257.pdf

Center for Sustainable Energy at Bronx Community College (January 2006). *New York City's Solar Energy Future*. Prepared for The City University of New York's Million Solar Roofs Initiative and Sustainable CUNY/Sustainable NY http://www.bcc.

cuny.edu/InstitutionalDevelopment/CSE/Documents/CUNY%20MSR%2   0-
%20Market%20for%20PV%20in%20NYC.pdf

Chambers, A.; Kline, D.M.; Vimmerstedt, L.; Diem, A.; Dismukes, D.; Mesyanzhinov, D.
(July 2005). "Comparison of Methods for Estimating the NOx Emission Impacts of Energy
Efficiency and Renewable Energy Projects: Shreveport, Louisiana Case Study." NREL TP-
710-37721. Golden, CO: National Renewable Energy Laboratoryhttp://
www.nrel.gov/docs/fy05osti/3772 1 .pdf

Del Chiaro, B.; Dutzik, T.; Vasavada, J.; (December, 2004). *The Economics of Solar Homes in
California*. Los Angeles, CA: Environment California Research and Policy Center
http://www.environmentcalifornia.org/uploads/J9/Kx/J9Kx6frBT7ZsGHGr3uUL9g/
Econ omics_of_Solar_Homes.pdf

Duke, R.; Williams, R.: Payne, A. (2001) "Accelerating Residential PV Expansion: Demand
Analysis for Competitive Electricity Markets." *Energy Policy* Vol.29; pp. 787 - 800.
http://www.princeton.edu/~cmi/research/Capture/Papers/accelerating.pdf

Energy and Environmental Economics, Inc.;Rocky Mountain Institute. (October 25, 2004)
*Methodology and Forecast of Long Term Avoided Costs for the Evaluation of
California Energy Efficiency Programs.* San Francisco, CA
http://www.ethree.com/cpuc/E3_Avoided_Costs_Final.pdf

Florida Public Service Commission; Florida Department of Environmental Protection (2002).
An Assessment for Renewable Energy Electric Generating Technologies for Florida.
http://www.psc.state.fl.us/industry/electric_gas/Renewable_Energy_Assessment.pdf

Hoff, T.E.; Perez, R.; Braun, G.; Kuhn, M.; Norris, B.; (March 17, 2006) *The Value of
Distributed Photovoltaics to Austin Energy and the City of Austin*. Napa, CA: Clean
Power Research LLC, http://www.austinenergy.com/About%20Us/ Newsroom/
Reports/PV-ValueReport.pdf ("Austin Energy Report")

Hoff, T.E. (2002) *Final Results Report with a Determination of Stacked Benefits of Both Utility-
Owned and Customer-Owned PV Systems*. Napa, CA: Clean Power Research, LLC
http://www.smud.org/pier/reports/S-034,%201.3.5.2,%2012-02,%20DEL(rev).pdf
("Smud Report")

Hoff, T.E.; Perez, R.; Margolis, R.M. (2005) *Maximizing the Value of Customer-Sited PV
Systems Using Storage and Controls*. Napa, CA: Clean Power Research, LLC
http://www.clean-power.com/research/customerPV/OutageProtection_ASES_2005.pdf

Hoff, T.E.; Perez, R.; Margolis, R.M. (November 9, 2004) *Increasing the Value of
Customer-Owned PV Systems Using Batteries*. Napa, CA: Clean Power Research,
LLC http://www.clean-power.com/research/customerPV/OutageProtection.pdf

Hoff, T.; Margolis, R. (June 6, 2005). *Moving Towards a More Comprehensive Framework
to Evaluate Distributed Photovoltaics*. Napa, CA: Clean Power Research, LLC and
Golden, CO: National Renewable Energy Laboratory http://www.clean-
power.com/research/customerPV/EvaluationFramework.pdf

Itron, Inc. (September 14, 2006).*CPUC Self-Generation Incentive Program, Preliminary Cost-
Effectiveness Evaluation Report.*Vancouver, WA http://www.itron.com/asset.asp?
path=assets/itr_001094.pdf

U.S. Department of Energy (no date available) Million Solar Roofs Initiative: Success Stories.
Residential Installation, On-Grid PV System, Glenn Dale, MD
http://www.nrel.gov/docs/gen/fy98/25718.pdf

Black & Veatch Corporation (May, 2003). *Minnesota Photovoltaic System Case Study: Residential Application.* Prepared for Iowa Department of Natural Resources

Navigant Consulting Inc., (February 12, 2006) *Distributed Generation and Distribution Planning: An Economic Analysis for the Massachusetts DG Collaborative*

Center for Sustainable Energy at Bronx Community College (January, 2007). *New York City's Solar Energy Future Part II. Solar Energy Policies and Barriers in New York City.* Prepared for The City University of New York's Million Solar Roofs Initiative.

Appendix VI PV Planner Assumptions for all Sectors http://www.bcc.cuny.edu/ institutionalDevelopment/cse/Documents/CUNY%20MSR%20   -%20Market%20for%20PV%20in%20NYC.pdf

New York Energy $mart. (November, 2004). "Darmstadt Overhead Doors." Fact sheet. http://www.powernaturally.org/About/documents/Darmstadt_SuccessStory.pdf

Perez, R., Letendre, S. (February 2006).*Understanding The Benefits of Dispersed Grid-Connected Photovoltaics. From Avoiding the Next Major Outage to Taming Wholesale Power Markets* http://www.sciencedirect.com

Perez, R.; Hoff, T.; Herig, C.; Shah, J. (May 2003). *Maximizing PV Peak Shaving with Solar Load Control. Validation of a Web-Based Economic Evaluation Tool* http://www.sciencedirect.com

San Diego Regional Energy Office. (no date available) Del Mar Fairgrounds SelfGen Case Study. http://www.sdenergy.org/uploads/SelfGen%20-%20Case%20Study%20-%20 DMFG.pdf

Rabl A.; Spadaro, J.V. (November 2000) *Public Health Impact of Air Pollution and Implications for the Energy System* http://arjournals.annualreviews.org/ doi/pdf/ 10.1146/annurev.energy.25.1.601

Smellof, E. (January 2005) *Quantifying the Benefits of Solar Power for California* http://www.votesolar.org/tools_QuantifyingSolar%27sBenefits.pdf ("The Vote Solar White Paper")

Watts, J.; (Fall 2006) "Solar Power Case Study: An Old Idea With New Economics." Solar Washington Newsletter http://www.solarwashington.org/newsletters/0609/Watts.pdf

Wisconsin Focus On Energy Case Study: Solar Energy in the City (2004). REN-2044- 1104. http://www.focusonenergy.com/data/common/dmsFiles/W_RS_MKCS_UW%20Gre en% 20case%20study.pdf

Wiser, R.; Mills, A.; Barbose, G.; Golove, W. (July 2007).*The Impact of Retail Rate Structures on the Economics of Commercial Photovoltaic Systems in California.*LBNL63019, Berkeley, CA: Lawrence Berkeley National Laboratory http://eetd.lbl.gov/ea/ ems/reports/63019.pdf

INDEX

D

E

T

U

V

W

Y